Healthcare Technology Management Systems

Healthcare Technology Management Systems

Toward a New Organizational Model for Health Services

Luis Vilcahuamán and Rossana Rivas

Reviewed by

Tobey Clark

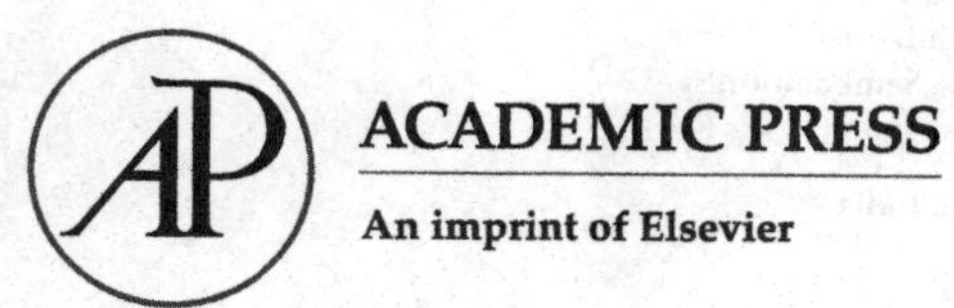

ACADEMIC PRESS
An imprint of Elsevier

Academic Press is an imprint of Elsevier
125 London Wall, London EC2Y 5AS, United Kingdom
525 B Street, Suite 1800, San Diego, CA 92101-4495, United States
50 Hampshire Street, 5th Floor, Cambridge, MA 02139, United States
The Boulevard, Langford Lane, Kidlington, Oxford OX5 1GB, United Kingdom

Copyright © 2017 Elsevier Inc. All rights reserved.

No part of this publication may be reproduced or transmitted in any form or by any means, electronic or mechanical, including photocopying, recording, or any information storage and retrieval system, without permission in writing from the publisher. Details on how to seek permission, further information about the Publisher's permissions policies and our arrangements with organizations such as the Copyright Clearance Center and the Copyright Licensing Agency, can be found at our website: www.elsevier.com/permissions.

This book and the individual contributions contained in it are protected under copyright by the Publisher (other than as may be noted herein).

Notices
Knowledge and best practice in this field are constantly changing. As new research and experience broaden our understanding, changes in research methods, professional practices, or medical treatment may become necessary.

Practitioners and researchers must always rely on their own experience and knowledge in evaluating and using any information, methods, compounds, or experiments described herein. In using such information or methods they should be mindful of their own safety and the safety of others, including parties for whom they have a professional responsibility.

To the fullest extent of the law, neither the Publisher nor the authors, contributors, or editors, assume any liability for any injury and/or damage to persons or property as a matter of products liability, negligence or otherwise, or from any use or operation of any methods, products, instructions, or ideas contained in the material herein.

Library of Congress Cataloging-in-Publication Data
A catalog record for this book is available from the Library of Congress

British Library Cataloguing-in-Publication Data
A catalogue record for this book is available from the British Library

ISBN: 978-0-12-811431-5

For Information on all Academic Press publications
visit our website at https://www.elsevier.com/books-and-journals

Publisher: Mara Conner
Acquisition Editor: Chris Katsaropoulos
Editorial Project Manager: Anna Valutkevich
Production Project Manager: Anusha Sambamoorthy
Cover Designer: Christian J. Bilbow

Typeset by MPS Limited, Chennai, India

Contents

Authors' Biographies

Luis Vilcahuamán has a wide range of experience in research, analysis, and instruction in Clinical Engineering, including a PhD from the University of Orleans, France and MSc from the Federal University of Santa Catarina, Brazil. He is currently a research professor at PUCP, and has been a research director of numerous projects in the Applied Science category of the Technology, Science and Innovation Fund FINyCT/Peru, as well as the National Fund for Science and Technology Development FONDECyT/Peru. He is director of Health Technopole CENGETS and director of the Biomedical Engineering Master Program at PUCP. He has been a consultant for healthcare technology systems for major national and international institutions such as the Pan American Health Organization, the World Bank, Developing Interamerican Bank, Social Security in Peru and San Salvador, Ministry of Health in Peru, as well as many Peruvian hospitals. He is a researcher in clinical engineering and rehabilitation engineering. Prof. Vilcahuamán is a member of Engineering in Medicine and Biology Society (EMBS-IEEE), International Federation for Medical and Biological Engineering (IFMBE), the American College of Clinical Engineering (ACCE), International Discussion Forum for Global Exchange: Infrastructure and Technology of Health Services (INFRATECH), and Regional Council of Biomedical Engineering for Latin America (CORAL). He participated in the World Congresses of the World Technopolis Association (WTA). In 2010, he received the American College of Clinical Engineering—ACCE and ORBIS International Award for having demonstrated significant improvements in Health Technology Management—HTM as part of a group of more than 23 countries around the world.

Rossana Rivas is a PhD candidate at the Ecole des Mines de Paris, France. She received MSc in Health and Social Organizations Management with a major in Management, Organization and Strategy at the Univ. Jean Moulin Lyon, France. She is a key leader and specialist in Health Technology Management, Clinical Engineering, Biomedical Engineering, Health Technology Assessment and Innovation particularly for Health Technology policy, leadership, planning, contracting, and establishing new programs, such as with MoH and EsSalud in Peru. Since 2006, she has collaborated to develop the Clinical Engineering Health Technology Management internship in partnership with the University of Vermont (UVM) that also supported Clinical Engineers in Colombia and Argentina. In 2015, she was a consultant with Pan American Health Organization (PAHO) for the Healthcare Technology Planning & Management online course aimed at 18 countries. She is a co-founder of Technopole CENGETS—an international known organization interacting with external entities nationally and internationally to improve the viability of the Health Sector through investigations, evaluations, studies, projects, and training. She is internationally recognized as indicated by her role at the 2015 World Congress of Medical Physics and Biomedical Engineering and her appointment as a collaborator to the Clinical Engineering Division of the International Federation for Medical and Biological Engineering—FMBE. At the Peruvian NIH she has focused on Health Technology Transfer, Maternal Health and Child development, and Heavy Metal Environmental Health; three of the NIH's top-priority areas in Peru today. She has been a consultant in healthcare technology management for the PAHO, the Canadian International Development Agency (CIDA), Resources International Group Ltd., USAID, and UNDP. She is a member of the American College of Clinical Engineering (ACCE), International Discussion Forum for Global Exchange: Infrastructure and Technology of Health Services (INFRATECH), the World Technology Association (WTA), Health Technology Assessment International (HTAi), Latin and Ibero American Association Technology Management (ALTEC), International Federation for Medical and Biological Engineering (IFMBE), Health Technology Task Group—International Union for Physical and Bio-Engineering Sciences in Medicine (IUPESM), Health Technology Assessment Discussions-Panamerican Health Organization (PAHO-HTA), Medical Device Announcements from the World Health Organization. From 2011 to 2015 she worked as a member of the Commission for the creation of the Biomedical Engineering Joint Career at UPCH and PUCP Universities. Rossana Rivas has received several awards, including the 2017 and 2015 International Federation of Medical Physics & Biomedical Engineering award; Science, Technology and Innovation International Internships; National Development Fund for Science, Technology & Technological Innovation—FONDECYT, awarded by the National Council for Science, Technology & Technological Innovation—CONCYTEC, Lima, Peru, 2015 and 2013, and the American College of Clinical Engineering, ACCE and ORBIS International Award for having demonstrated significant improvements in Health Technology Management—HTM as part of a group of more than 23 countries around the world in 2010.

Introduction

Problems are part of solutions, it is said. What is going on in hospitals?

The book is aimed toward technology decision makers, stakeholders and users in hospitals, institutions of research, health regulatory agencies, and other related organizations. In this regard medical staff, engineers, nurses, care technicians, managers, researchers, policy-makers, head of clinical services, and students are considered as the audience. We think also that the book will provide students and professionals a practical approach to improve the technology status in hospitals and other healthcare organizations.

We observe that the status and the utilization of technology in hospitals change according to the surrounding environment. Certainly effective regulations, appropriate budgets, and an adequate organizational culture result in a better response from the health organization but in general all of them have distinctive problems. In developing countries problems are related to low-level medical devices operability, undetected high risks resulting from the use of technology, high costs and the disarticulated work that characterizes the labor of health staff, administrative staff, the technical team, and the engineers.

Biomedical Engineering—BME brings a distinctive impact to medicine and health, in particular Clinical Engineering—CE which focuses on the improvement of technology status in hospitals, BME and CE are relevant change-drivers, but in general they find barriers for the implementation expected. Next to the observation of the current status of technology it is obvious that something is not working or something is missing. A trusted health service, with clinical effectiveness and reasonable costs, requires a strategy. The authors worked on this book to provide insights for new ways to define the strategy. For this purpose, we include a holistic concept of what technology is, the integration, the inter- and multi-disciplinary approach, the network of communication between the stakeholders and, last but not least, the approach that the patient is the priority person and job recipient of health staff, administrative, technicians, and engineers.

Raising the target: In hospital functioning with appropriate technology, are we talking only about medical equipment? Are purchase, installation, and maintenance activities all we need? A skillful engineer is all which is required to accomplish the target? Is it needed to hire an engineer? or it is enough a technician. What is the suitable coordination between the engineer or technician and the medical and administrative staff? These and other questions are unresolved in the hospital, so we think the best is to start at the beginning.

Technology and Technic are both relevant concepts but have a different meaning; for hospitals, the best is to talk about Technological level capacity and Technical level capacity. See the difference as follows:

Technology [Webster Dictionary]: (1) The practical application of science to commerce or Industry (syn. Engineering). (2) The discipline dealing with the art or science of applying scientific knowledge to practical problems.

Technique [Webster Dictionary]: (1) Practical method or art applied to some particular task. (2) Skillfulness in the command of fundamentals deriving from practice and familiarity (syn. Proficiency, facility).

In order to have a hospital with *technological capacity*, it is necessary to have the scientific knowledge applied to the practical problems. On the other hand, to have the *technical capacity* is required to have the skill achieved through fundamental and daily practice. Both are basic and essential, also, the proper balance between them is critical. The technological complexity of the hospital is increasing and requires professionals with a capacity for analysis and synthesis, that is, they have scientific knowledge to solve the problems, be they engineers, physicists, or even administrators or architects. In this sense, we consider the scientific knowledge a basic capacity in the hospital regardless of the size of the hospital. Since it is impossible for the professional to be an expert in all technologies, the natural tendency is to move toward specialization, as is the case with specialization in the various clinical services. The current complexity of hospital technology demands that experts in the hospital field be consulted and at this point it is practically inappropriate to consider an engineer with a traditional education away from hospitals, it is true that every engineer will be able to learn and adapt in some way, but it is also true that particular training is required to master the technologies currently in use in hospitals such as biomedical engineers and clinical engineers.

HEALTH TECHNOLOGY

According to the World Health Organization [1], when referring to Health Technology the definition must include clinical technologies (medical procedures, medical devices, drugs, and medical materials); support technologies (infrastructure and hospital systems, energy systems, information systems and communication, and the organization itself); also technologies for community health should be included: prevention technologies, protection, and promotion; and even technologies for environmental health. Technology has reached high levels of complexity, sophistication and today it is essential to be able to provide health services with the expected level of quality [2]. Complexity and risk related to the investment determine having an approach of new organizational structures and management, both key to deciding issues effectively taking into account the high costs of acquisition and operation, risk control, clinical effectiveness and efficiency in the use of technological resources; all of which makes a sustainable health system whether in a developed or developing country. Consider, as we did in the 1960s that health technology is only a matter of repairing medical equipment is obsolete and inadequate. Technological resources are more than ever, capital assets of high economic value, essential in clinical service and require specialized decisions focusing on the benefits of investment. Nowadays, it is not enough to have a medical devices operative, what really serves to medical staff and especially what serves patients are functional clinical environments, absolutely everything technological must work to the point of being effective for the clinical procedures. This is one of the milestones of this book; to achieve a functional clinical environment as a result of integrated and specialized work.

THE TECHNOLOGY USER AND DESIGNER

The lack of hospital technology specialists determines that in many places the users, called healthcare staff, require to complement the designer approaches and strategies. This is very common, even useful if developed carefully, but it is also harmful often. The following analogy will help us to understand this: the captain of the airplane is in charge of the passengers during the flight, in a similar way it is the doctor who is in charge of his patients. On the plane, the captain is an expert staff in all maneuvers and forecasts in order to carry passengers safe and on time to their destination, he knows in depth the operation and limitations of the airplane, he may even review the functionality of the airplane and propose improvements. However it is not the captain who designed the plane. They were specialists who designed the plane probably not known to the captain nor the passengers. So why assume that the healthcare staff should provide all the information required to design a clinical service? The information and opinion provided by the healthcare staff is important, but as in the case of the airplane, designing a hospital requires having expert designers with different range and level of competence. The user of technology has a different field of action with regard to technology designers and both must interact. On the other hand, health system must have technology designers. They may be biomedical engineers, clinical engineers, hospital engineers, health managers, architects of hospitals, medical physicists, and other specialists of the health sector, which in the 21st century is a sector by nature interdisciplinary [2]. It is clear that today in many places this expertise does not exist or there are insufficient skilled health professionals in technology.

ENGINEERS IN THE HOSPITAL

To what extent can the engineers intervene in a hospital? Traditionally hospitals in developing countries, health organizations call for engineers with conventional education: electrical, electronic, mechanical, industrial engineers, etc. By observing the curriculum, there is few or no courses related to hospitals, hopefully there are some additional courses in biomedical. We think that the traditional education of engineering is not enough to work in hospitals. How can the engineers work adequately with the previously described types of technology without specialized education? The conventional engineers, untrained in hospitals, tend to address only the maintenance and neglects the other substantive interventions such as the design of clinical services, support for the development or implementation of clinical procedures, analysis of events related to safety, assessing cost-effectiveness of the technologies used, technology planning, design of technology strategies, among many other possible and necessary interventions that this book describes. An essential cross-cutting issue should then be observed: health sector demands professionals with strong background in biomedical and hospital technology, based on physiology, mathematics, physics, chemistry, and biology of the human body, in addition to engineers being able to apply engineering sciences to the clinical aspects of the hospital.

In developing countries, hospitals often have a distant organizational structure related to the integrated technology management. Consequently, the engineer is not motivated to develop a career in hospitals. The resulting inefficiency and lack of clinical effectiveness, both direct consequences of the situation described above, will generate possibly much higher unnecessary costs to the investment that would be realized if the hospital had hired the right people. The negative scenario then is against one of the relevant objectives of the hospital: raising the functionality of the technology in a cost-effective way. The paradox here described is still poorly understood and explains why hospitals lack economic resources due to technological decisions (see Chapter 4: Health Technology Planning and Acquisition) [3], and it also lets us understand the reason for resistance in hospitals to incorporate expert engineers.

We think a key aspect required to achieve the expected objectives is that the modern engineer for hospitals should take in count, either a biomedical engineer or clinical engineer or other similar professional interacting not only with the healthcare staff, but keep also an open mind to other disciplines such as architecture, economics, law, and administration. To do this, it is required to think of a new structure of hospital organization and its processes, a hospital structure in which all the aspects of technology work in a coordinated and functional way to the benefit of the patient and the health professionals. This is another milestone of this book.

ETHICS IN HEALTH ENGINEERING

Related to this book, we cannot fail to mention our commitment to ethical principles concerning Health Engineering [4,5]: beneficence (benefiting patients), nonmaleficence (doing not harm), patient autonomy (the right to choose or refuse treatment), justice (the equitable allocation of scarce health resources), dignity (dignified treatment of patients), confidentiality (of medical information), informed consent (consent to treatment based on a proper understanding of the facts), and human enhancement (design to enhance healthy human traits beyond a normal level).

The purpose of the book then is to provide a model and its components to implement an effective Healthcare Technology Management (HTM) system in hospitals, and reflect on the need to rethink the hospital organization for decision-making processes related to technology. Current models of management and organization of technology in hospitals have evolved over the last 40 or 60 years ago, according to totally different circumstances than now, therefore, they all have a "factory default." Our proposal is that in the context of new technologies it is not enough to update the obsolete model, but requires a re-engineering of management and organization to achieve adequate levels of clinical effectiveness, efficiency, safety, and quality that users expect of the technology used in hospitals. Many of the current premises on good practices in HTM provided by specialized institutions on health are impracticable due to the lack of human resources, responsibility, and adequate procedures for the implementation of the proposed processes.

The book is aimed at decision makers, stakeholders and users of technology in hospitals, research institutions, health regulatory agencies, and other organizations related. In this regard, three aspects are relevant in the book: (1) the focus from a "field perspective" in health technology and the "holistic approach," including the academia background and research; the perspective then is not general but integral; (2) the inclusion of the perspectives, knowledge, and best practices of expert global organizations; and (3) the two authors' experience on exchanging with health sector stakeholders from developed and developing countries over a significant number of years to the present.

Finally, we think that the current problem regarding the status of medical devices can be improved in a viable way. The contribution of many experts and researchers in many different parts of the world over recent years has established advances and open doors. The following steps may be oriented to consolidate a better organization and processes for proper management, not only with respect to medical devices, but referred to the technological environment in hospitals. The solution designed in this context will largely exceed the original problem, giving better opportunities for patients and for everyone in general.

References

[1] Pan American Health Organization (PAHO), Developing health technology assessment in Latin America and the Caribbean, 4th ed. Washington, D.C.: PAHO; 2000.

[2] Vilcahuamán L, Rivas R. Ingeniería Clínica y Gestión de Tecnologías en Salud: Avances y Propuestas. Lima: CENGETS PUCP; 2006.

[3] WHO. The world health report: health systems financing: the path to universal coverage. Geneva, Switzerland: World Health Organization; 2010.

[4] Fielder J. Biomedical engineering ethics. San Rafael, California: Morgan & Claypool; 2007.

[5] Varello D. Biomedical ethics for engineers: ethics and decision making in biomedical and biosystem engineering. San Diego: Academic Press; 2007.

CHAPTER 1

Healthcare Technology Management (HTM) & Healthcare Technology Assessment (HTA)

The task is to maximize the benefits and minimize the risks of technology in hospitals.

HEALTHCARE TECHNOLOGY MANAGEMENT—HTM

The technology has been used in hospitals since their inception. However, the study of the impacts of technology on people's health, the cost analysis, and management of its operation has not been fully structured and systemized in hospitals. Two trends that have been developing in recent years are healthcare technology management (HTM) and clinical engineering (CE), and healthcare technology assessment. The last topic will be developed in the second part of this chapter.

1.1 HEALTHCARE TECHNOLOGY MANAGEMENT AND CLINICAL ENGINEERING

It is natural to think that at the beginning of the concept of the hospital, especially when the first electrical medical devices appeared, the need for engineers and technicians to handle its operation was apparent. With the increasing number of medical devices and technological complexity, an area of engineering expertise was created, often associated with the electrical engineer or electronic engineer. This new professional field later formed the CE and biomedical engineering. The first one closely associated to the work within hospitals and the second one with a greater emphasis on research and health technology design. CE can be considered as a specialty in biomedical engineering, although in some places the CE begins to take its own course. However, one could say that what most distinguishes between a biomedical engineer and a clinical engineer and the more traditional engineers are their training in physiology making use of a thorough study of biology, physics,

Healthcare Technology Management Systems. DOI: http://dx.doi.org/10.1016/B978-0-12-811431-5.00001-1
© 2017 Elsevier Inc. All rights reserved.

chemistry, and mathematics. The traditional training of doctors in clinical services bypasses this approach, so it could be argued that the clinical engineer and biomedical engineer are sufficiently different from other engineers and doctors themselves. The American College of Clinical Engineering (ACCE) defines a clinical engineer as: "A Clinical Engineer is a professional who supports and advances patient care by applying engineering and managerial skills to healthcare technology," definition since 1992.

On the other hand, the HTM is a specialized form of management. Given its approach to technology in health, it would be considered doubly specialized. This discipline nourished in the fields of management and administration has been developing quickly and in this case, guides much of the work of clinical engineers. HTM is defined as "the systematic process in which qualified healthcare professionals, typically clinical engineers, in partnership with other health-care leaders, plan for and manage health technology assets to the highest quality care at the best cost." Clinical Engineering Handbook, J. Dyro, 2004.

This is where it becomes necessary to remember the concept of health technology referred to in both definitions. We tend to associate the technology only with medical devices and sometimes the technical aspects as was already explained in the introductory chapter. However health technology is much broader and goes beyond the scope of just a specialist. Fig. 1.1 shows the diversity of health technology.

It can be concluded first that more than one type of specialist is required to address this diversity of health technologies. It must be considered that a hospital contains all types of technology. From the point of view of a hospital and the patient, if we only focus on medical devices, we will be doing a partial job, and even more so if the clinical service is expected to be functional, then it is required that all types of technology are functional. Under these assumptions, the HTM involves all these types of technology, but what should be the scope of work of the clinical engineer? and who is responsible for each type of technology in a hospital?

Various professional profiles are needed to address the technology of a hospital with all having sufficient training in basic sciences and HTM, in addition to training in physiology and clinical aspects associated to the clinical service. In the case of clinical engineer, the matter is interesting because you can define different nuances depending on the objectives to be achieved. In solid and well-structured hospital organization, a clinical engineer with guidance related to medical devices is feasible because other professionals will cover other types of technology. However in organizations with weak organizational structure, as is common in developing countries, a clinical engineer is required with a capacity to conduct various types of technology at the

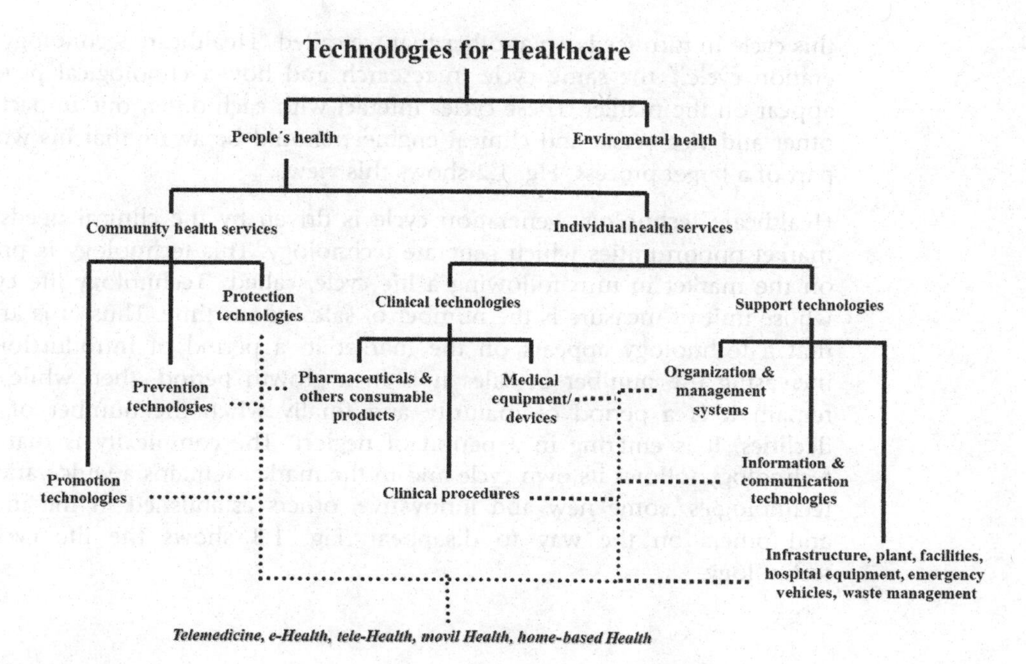

FIGURE 1.1
Types of healthcare technology to consider in HTM & CE. *Modified of Developing Health Technology Assessment in Latin America and the Caribbean—PAHO/WHO.*

same time. For example a clinical engineer with capacity for management of medical devices and with sufficient knowledge to define and evaluate infrastructure requirements, ICT and management systems, and organization would be very useful in in this situation. Many hospitals have a research mission, in this case the clinical engineer should be involved with the development and testing of clinical procedures, together with medical professionals. This fact raises a somewhat different profile of clinical engineer than usual, which may include postgraduate, but will be appropriate for the current reality.

From the point of view of responsibilities to a technologically functional hospital, this is explained from the technology life cycle in a hospital which we have called "Healthcare technology application cycle." Processes to incorporate technology, use them and then replace or discard, follow a cycle in time, as happens in many other cycles in nature and organizations. It contains well-established stages from which the tasks, functions, roles, and therefore the positions and responsibilities of technology in a hospital are defined. Before addressing the development of the healthcare technology application cycle, we consider it important to present a more holistic view, as this cycle in turn feeds on another that we called "Healthcare technology generation cycle," the same cycle in research and how technological products appear on the market. These cycles interact with each other, one impacts the other and vice versa, and clinical engineer should be aware that his work is part of a larger process. Fig. 1.2 shows this view.

Healthcare technology generation cycle is driven by the clinical needs and market opportunities which generate technology. This technology is present on the market in turn following a life cycle, called "Technology life cycle," whose unit of measure is the number of sales versus time. Thus, it is known that a technology appears on the market in a period of introduction, by increasing the number of sales it is in a growth period, then while sales remain it is a period of maturity and finally when the number of sales declines, it is entering in a period of neglect. The complexity is that each technology follows its own cycle and in the market remains a wide variety of technologies, some new and innovative, others established in the market, and others on the way to disappear. Fig. 1.3 shows the life cycle of technology.

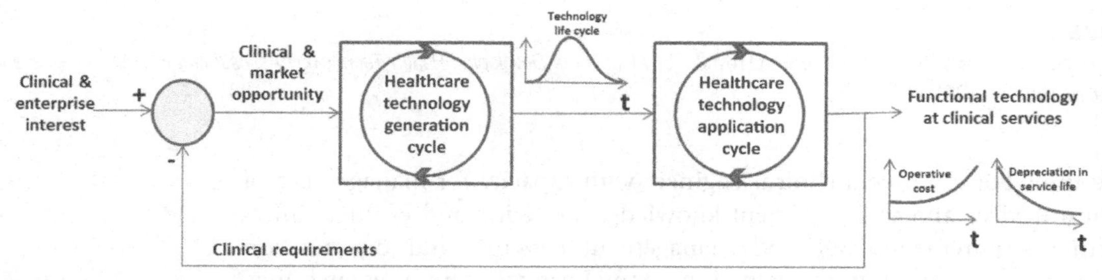

FIGURE 1.2
Holistic view of technology generation cycle and technology application cycle in healthcare.

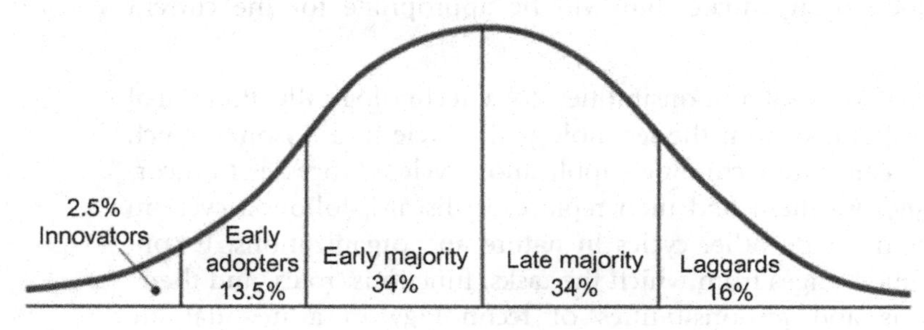

FIGURE 1.3
Typical technology life cycle. *Everett Rogers Technology Adoption Lifecycle model.*

Naturally the technology that reaches the hospital comes from market and it is therefore subject to variations thereof. When deciding what technology to buy, it will require making an analysis of technology life cycle. For example, obsolete technologies may inadvertently be the option to buy, but neither buying the latest technology would not be the ideal choice until to have opinion of other users. It is also necessary to analyze the technology life cycle to determine the best technology for our clinical needs considering the current reality of the hospital.

From the point of view of the hospital, the technology follows stages described by the Healthcare technology application cycle and a good management of this cycle will determine the functionality of the technology in the hospital. It is also from the hospital and the health sector in general that clinical requirements are identified, which is useful for other entities associated with health and industry who are interested in developing technology. These factors determine the generation of new and better technology according to the opportunities of the moment, and this is how technology generation and application of these in the hospital are part of a larger cycle.

Another complexity is added to take into account the types of technology. It can be clearer if we consider medical devices as one of the types of technology, but also clinical procedures follow a generation cycle, an application cycle, and a life cycle on the market or professional environment. Also, the hospital infrastructure and equipment follow these cycles although in this case their times are much greater, a hospital can have a lifetime of 40 or 50 or more years. Therefore, diversity and quantity of technology creates a complex environment for HTM, but it is precisely this environment that also provides flexibility and opportunities to better meet the clinical needs of the hospital.

1.2 HEALTHCARE TECHNOLOGY GENERATION CYCLE

The hospital is the recipient of technology but this comes from a series of previous processes which together have the characteristics of a cycle that follow a sequence. At the same time, feedback to itself to continuously generate technologies is what we see in the market. Fig. 1.4 presents a model, although we know there are several ways to explain it.

Research and development is the primary activity for technology generation through prototypes, patents, clinical trials, and publications. At this stage, researchers from universities, industry, or research centers participate; in many places, the key is the interaction with industry, government, and users. Also, the availability of funds is key and they tend to come from the industry, the government itself, and other entities such as nonprofit or NGO.

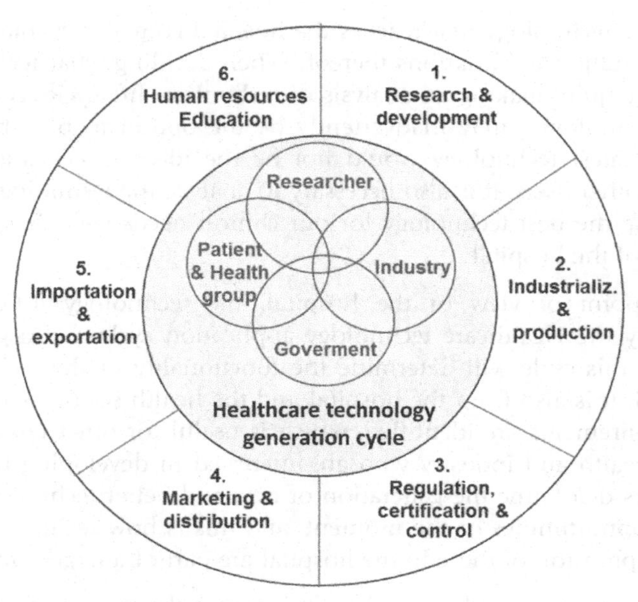

FIGURE 1.4
Healthcare technology generation cycle (market point of view).

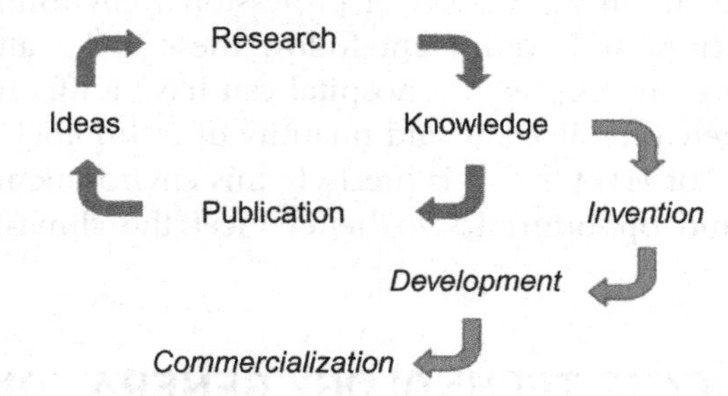

FIGURE 1.5
Healthcare technology transfer process.

Again we must remember here the diversity of possible technologies to develop which is explained in Fig. 1.1. For the results of research to be developed into market products, other no less complex processes are required, such as turning the prototype into a product with commercial value. The process of technology transfer (Fig. 1.5) is explained in this way according to the NIH Technology Transfer Program [1]: "new ideas prompt research, research yields knowledge, knowledge is shared through scientific and health-related publications, giving rise to new ideas and further research. Many of these ideas advance fundamental understandings of biology leading either to new strategies for the treatment or prevention of disease or to public and private sector development of medical products or services. Some of these ideas constitute inventions that lead more directly to products of interest to the public consumer and 'spin out' of the research cycle to enter a development phase where they are further refined, tested, and commercialized - moving into the marketplace as useful products and services to benefit public health."

On the other hand, every product on the market must comply with regulations imposed by governments to protect their citizens and compliance with the administrative requirements. The regulations cover the stages of presales (design and production), marketing, and after-sales. We will expand more on this topic in the next chapter. Also, for a technological product to reach the user, it is required for the company to set marketing and distribution strategies, in addition to import and export as appropriate. It is also important to set precise strategies to publicize the products and their benefits to users through events, publications, studies, specific trainings, etc. In a sense of virtuous circle, all actors such as users, researchers, government, and industry contribute to a feedback for new and better technologies.

The healthcare technology generation cycle is not natural; it is something that countries should build on their own in order to have a biomedical industry, so it is critical to have appropriate policies to strengthen them. It should also be noted that this cycle is the result of a collective and participatory work of various sectors of society. The researchers, industry, or government can't build it alone, and hospitals, as technology users and generators of clinical requirements, should not be neglected.

1.3 HEALTHCARE TECHNOLOGY APPLICATION CYCLE

The incorporation and use of technology in a hospital is explained by the healthcare technology generation cycle. The technology comes from the market and has a great diversity of not only types, but alternatives, brands, models, and all in a very dynamic environment and competition. It can be said that this cycle has a certain implicit nature with stages well established, to follow a fixed sequence, such as children and youth precedes it to adulthood. Another feature is that this is a closed loop cycle different from the life cycles where we have a beginning and an end. It means that technology is incorporated into the hospital to meet a clinical need. After the lifetime or service time in the hospital, the technology is discarded, however the need remains and a new technology will replace it, restarting the cycle again.

The ability we have to manage processes throughout the healthcare technology application cycle will determine the level of functionality of the technology used in the hospital. This functionality is described using five

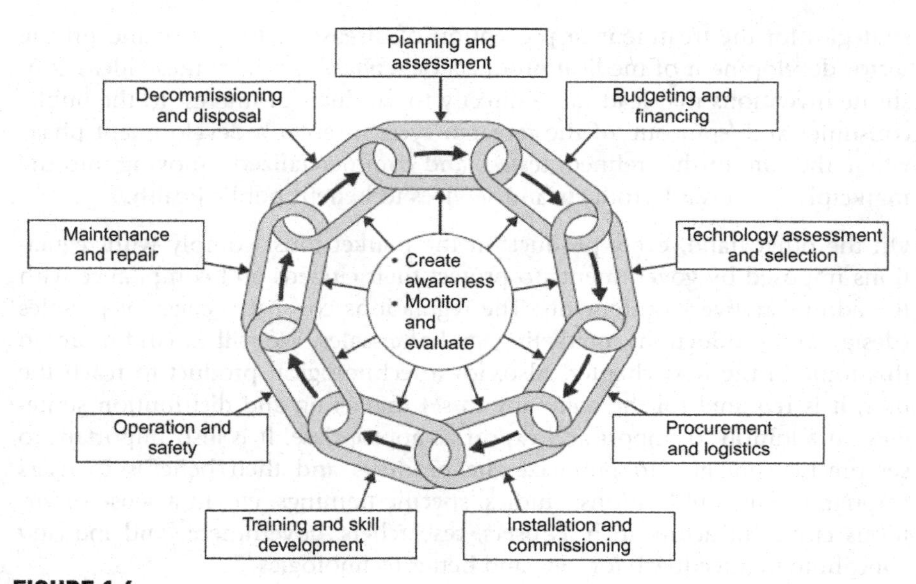

FIGURE 1.6

Healthcare technology application cycle in a hospital. *Guide 1: How organize a system of healthcare technology management. 'How to manage'. Series for healthcare technology. WHO, Ziken International Consultants UK, PAHO, GTZ, DFID [2].*

motivational factors: greater clinical effectiveness, greater efficiency, increased safety in the use of technology, cost reduction, and better quality of care in health services. Reliability and availability of technology are important cross-cutting issues that are included along these motivational factors. So, observing the details of these five factors can be known as the actual state of technology in a hospital. Fig. 1.6 shows a sample of healthcare technology application cycle.

Normally the authors present the technology application cycle oriented to medical equipment, and once again we must remember that each type of technology presented in Fig. 1.1 will follow an application cycle something different, for example in the case of medicines, some parts of the cycle in Fig. 1.6 should be changed, storage instead of installation, drug delivery instead of operation, and storage monitoring instead of maintenance. Also, if it will be a prevention technology such as vaccines, it should be considered the "cold chain" and discarding in case of it deteriorates or if the lifetime is exceeded. We propose the exercise of developing a technology application cycle for each types of healthcare technology. It is also necessary to say that the apparent simplicity of the cycle is not so much considering the amount of existing technologies in the hospital. It can be 1000, 2000, or 3000 medical devices in a hospital. Therefore each of these will be in any stage of the cycle, which creates greater complexity for management.

One aspect rarely analyzed is the identification of the person or entity responsible for each stage. It happens that in many hospitals, especially in developing countries, steps are being handled in different units of the hospital or no one is identified as being responsible for some stage. This breaks the integrity and interaction between entities, resulting in the functionality of technology in clinical services being poor. In many cases the planning and budget are managed by the unit of general services; logistics is another unit; the engineering department is responsible for the installation, operation, and maintenance, but not safety, for which there is a committee and so with this dispersion of responsibilities, nobody in particular is responsible for the functionality of clinical services, with the exception of general director. The planning of a better or a new organization requires beginning from these fundamental aspects to ensure that the healthcare technology application cycle has a coherent management between stages with an identified responsible entity for each. A responsible entity should be identified for the functionality of the all hospital technology. It is in this sense that we understand the concept of integrated healthcare technology management system.

The beginning stage of the cycle is the assessment of the clinical needs and planning of technological resources. These directly impact to the care services and therefore patients. This work is one of the most evident of the need for clinical engineers, because it requires knowledge of biomedical technology and also to understand clinical procedures and physiological consequences to weight the relevance. The list of needs is usually long because there are a lot of requirements, typically is not possible to have everything, so it is necessary to prioritize according to institutional strategic plans, annual operating plans, or particular goals identified. Also it requires evaluating a void which takes place with recently deregistered technology which requires replacement. Clinical needs that identify technology requirements can be certified as does the CENETEC in Mexico [3] and is evidence of a professional work. Once again we must consider the different types of technology to achieve the functionality of clinical service which is what in the end serves the health-care user and the patient. The clinical requirement may not only be the medical equipment and supplies, as the technology may require specific facilities. The requirement can be any of the types of technology referred to in Fig. 1.1, such as alterations in the information system, changes in the staffing or processes in the organization, clinical procedure change, or adaptation of the infrastructure. These and many other similar aspects should be analyzed, either calling for clinical engineers with different expertise or a clinical engineer with a particular profile. Interaction with other professionals is equally important. Communication and positive interactions with doctors, nurses, architects, administrators, or other engineers and technicians always will be appropriate to understand the functionality that we all want to reach.

For planning, it must determine the financial resources available to program acquisitions. Whether a public or private hospital, organizations follow procedures previously established for the management of funds. For example according to the expenditure, we have different alternatives: direct purchase, local tender, project investment, international tender, or others. While administrative support is required, the aim of purchase is technology, so the participation of clinical engineers is required. Interaction with suppliers is important so that institutional policies and good practices are required and are of benefit to the hospital. The ability to negotiate and validate an impartial information is very important, and the information products of ECRI, WHO, IMDRF, CENETEC, etc., are valuable. In case of calls for technology providers, the hospital usually forms a purchasing committee to evaluate the proposals. This work requires the expertise of clinical engineers to determine the best options that meet the clinical requirements. In some hospitals, the old model is followed which calls for engineers, architects, or other professionals with expertise out of a hospital, however it should be noted that technology in current health means having specialized professionals such as clinical engineers, medical physicists, architects for hospitals, and so on, with training in health technology. In the next chapters we will expand much more specific planning and acquiring of technology.

After buying technology, the hospital receives the products and they must be installed. The installation must also be planned as were technology purchases, since they entail costs and a series of preparatory work. Therefore it is necessary to emphasize that planning is not only evaluation and programming, it also involves integrated design features and provides space requirements, energy, installation, operating procedures, training, operating costs, logistical procedures, staffing, etc. In others words it is necessary to have a master plan or overall design of clinical service to provide all necessary considerations when making decisions.

These and other considerations mentioned above suggest that it is required for a single large unit that handles all stages of the healthcare technology application cycle in order to properly attend the tasks and interactions. This approach is more valid in those hospitals in developing countries that are considering a new or a better form of organization because they currently have one that is lagging far behind of the new principles and criteria of healthcare technology management and CE. Also is necessary to say that there are methods of healthcare technology assessment to be applied across the board in the different stages of the healthcare technology application cycle. This topic will be developed in the following pages of this chapter.

When the technology is already incorporated in the hospital, it begins the longest stage in the time that corresponds to its use by users such as the physician and allied health professionals. Many years ago the operation of medical devices was the primary purpose of this stage; however under the new premises appear other no less important aspects such as safety and ensuring the functionality of clinical services that far surpass the operation of only medical devices. For healthcare medical staff, functionality of all types of technology in clinical service matters, such as described in Fig. 1.1, and for clinical work all types of technology must be operational: energy systems, infrastructure, hospital equipment, information systems, training in clinical procedures, drugs, medical devices, protective devices, prevention devices, and with clarity in the organization at different levels of action regarding performing clinical procedures. All this makes a functional clinical service from the technological aspect. In this sense, in the next chapters we will describe the processes, functions, and their variables to achieve this goal.

The operability of technology, understood as the different types of technology, is a complex task due to diversity and quantity involved, proper management requires adequate computer support. It seems that in these times the manual management of documentation is also impossible and inefficient. In the case of equipment, energy and infrastructure are required to implement software systems for corrective and preventive maintenance and as well as inspections, which are proactive actions to check the functionality at the site of operation. Other technologies such as clinical procedures and organizational systems require evaluating the results, analyzing the situational status, evaluating the protocols and processes, performing measurements, acquiring data, among other actions to determine the operability and make improvements or adjustments if it is necessary.

The operability understood in this way must be complemented with security aspects of the technology in use. Every technology has implicit hazards that should be minimized to protect people. This issue leads to the need for risk management. At this point, the understanding of the operating principles of technology is essential, so it is evident to insist on rigorous training in the exact sciences (mathematics, physics, and chemistry) and life sciences for clinical engineers and other professionals such as medical physicists and biochemists. A functional technology must be safe for the patient, users, and people in general. For that to occur, instrument measurement capacity, data interpretation, and evaluation of impacts on people is required. In addition to compliance with the regulations, attention to technical standards and safety regulations is required. In the same way, risk management involves technology monitoring and recording of adverse events as a means to learn from mistakes, prevent recurrence, and improve the quality of the technology used in the hospital. Biomedical metrology also is part of the task of the clinical engineer to ensure adequate accuracy of the measurements used for diagnosis and treatment in the hospital. The measurement error is one of the major problems in many hospitals that lead to false positives and false negatives, while generating an unnecessary waste and reducing the quality of health services.

Finally, if technology is past its lifetime or service time in the hospital, it must be transferred, discarded, renewed, or replaced by other more effective or more efficient or safer technology. Hospitals should have clear policies for this procedure and have the information regarding cost, operability, safety, and effectiveness of the results of these technologies in order to make appropriate decisions.

1.4 PROCESSES AND ORGANIZATIONAL ELEMENTS FOR A HEALTHCARE TECHNOLOGY MANAGEMENT SYSTEM

After presenting the healthcare technology application cycle in hospitals, one wonders how this cycle will be implemented in a hospital? Naturally one of the options is to improve or adapt what you already have in the organization. Another alternative is to implement a new organizational structure specializing in technology management. Our particular experience makes us think that the second option is more viable and better prognosis in the long term. This is a critical and probably moot point, surely additional validations and larger studies will be required to confirm it, however we believe that what has been said here and with the development of the following chapters the value of this position will be demonstrated.

In clinical services there are interactions between health professionals, administration, and technology, in all types as described in Fig. 1.1. The goal is to maximize the benefits and minimize the risks of technology in hospitals. To this the healthcare technology application cycle explains the stage in its own nature by which the technologies are incorporated and after use, they are discarded or renewed. For better description, Fig. 1.7 shows processes and stratums that support the healthcare technology application cycle, which in turn allows that a clinical service technology be functional for the hospital. Proper handling of the healthcare technology application cycle is based on four main processes (pillars). These are the planning, acquisition, property management, and risk management. Naturally, little could be done in an organization if there were no policies in technology. In fact it is known that the lack of policies on technology in developing countries is one of the main factors that prevent incorporation of many of the premises of CE. Health ministries are organized to provide clinical services and manage their

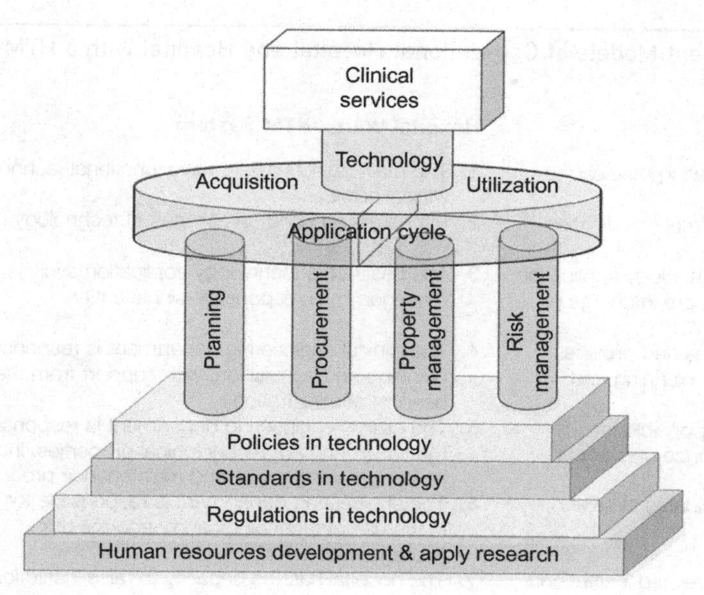

FIGURE 1.7
Processes and organizational requirements for functional technology in clinical services. *Modified of WHO and Tobey Clark-UVM.*

administration, but the technology management shows a big weakness. In addition, compliance with standards and regulations are cross-cutting issues in all processes to ensure the safety and clinical effectiveness. Just the lack of policies on technology makes the application of many standards does not take into account.

We have also considered two additional fundamental processes to make an organization of this type viable: The development of human resources in technology and applied research in biomedical technology. It will be appropriate to have in the hospital with professional clinical engineer or biomedical engineers trained in universities, and still it will be necessary to complement this training with the experience and the developments in the hospital, as well as upgrades and specializations that the hospital needs and the career itself requires. It is also necessary for the health professionals and technical staff to train in healthcare technology, and in this sense to project to the hospital network attached through training. These training requirements and continuous training can be managed through specific and specialized processes. On the other hand, in a context of evolution and continuous changes as they occur in hospitals knowing the dynamism of technology and because of the commitment to quality improvement, it is necessary to have the ability to study, analyze, design, and implement new or better processes, methods, and procedures. In addition, to improve and even to innovate in

Table 1.1 Comparing Management Models of Conventional Hospital and Hospital with a HTM System

Conventional Hospital	Hospital With a HTM System
1. The main objective is to have operating medical devices	**1.** The main objective is to have functional technology in clinical services
2. Efforts are focused on maintaining medical devices	**2.** Efforts are focused on all types of technology (see Fig. 1.1)
3. In the best case the healthcare technology application cycle is handled, but different units are in charge of the stages	**3.** The healthcare technology application cycle is managed by a responsible single entity
4. The engineering department advises and provides opinion to the responsible units for planning and acquirement	**4.** The clinical engineering department is responsible for planning and acquisition, with support from the hospital administration
5. The engineering department is responsible for corrective and preventive maintenance of medical devices	**5.** The clinical engineering department is responsible for the functionality of technological properties, including property management and maintenance programs
6. The engineering department participates in safety committee of the hospital	**6.** The engineering department is responsible for safety in the use of technology and performs risk management
7. The hospital have few technology-related institutional policies	**7.** The hospital has the capacity to raise institutional policies on technology
8. Compliance with regulations is shared with other units of the hospital. The compliance is incomplete due to the lack of policies	**8.** There are explicit processes for compliance with standards and regulations
9. The training of human resources is limited and does not include the medical assistance staff	**9.** The training of human resources is for all hospital staff and with explicit processes
10. The engineering department staff does not usually participate in research	**10.** The personnel of clinical engineering department have the capacity to participate in research together with health professionals

the technologies used in the hospital leads us to believe that the hospital must have the capacity for applied research in health technology. Without this ability the hospital tends to stagnate which is directly counterproductive for patients and users of technology.

We will make a brief comparative analysis with the typical organization of a hospital in a developing country and a hospital with a HTM system (Table 1.1).

HEALTH TECHNOLOGY ASSESSMENT—HTA

1.5 DEFINITION

According to the European Network for Health Technology Assessment—EUnetHTA [4], *Health Technology Assessment—HTA* is a multidisciplinary process that summarizes information about the medical, social, economic, and ethical issues related to the use of a health technology in a systematic, transparent, unbiased, robust manner. Its aim is to inform the formulation of safe, effective, health policies that are patient focused and seek to achieve best value.

To have the capacity to work as a team, to research and exchange information from a multidisciplinary perspective is relevant to obtain the results expected according to EUnetHTA.

HTA's purposes stated by International Network of Agencies for Health Technology Assessment—INAHTA [5] are as follows:

- Primary purpose: To inform decisions relating to national, regional, or local healthcare systems. Such decisions may relate to the procurement, funding, or appropriate use of health technologies and also to disinvestment in obsolete or ineffective technologies.
- Secondary purpose: To contribute to global knowledge on assessment of specific technologies—a library function. HTA provides source material for other research, guidelines, etc.

Outputs expected from the implementation of HTA's: (1) Products related to HTA as guidelines, economic assessments, reports, etc.; (2) Decision Making which involves definition and selection of the possible alternatives aimed on solving problems.

The complexity and gradually higher costs involved in health sector activities demand the implementation of HTA in developed and developing countries. Despite the financial, logistic, and other type of investment, we hopefully agree on having HTA's capacities in health sector as a key factor to provide value to the patients, the families, and the society in general.

According to World Health Organization—WHO [6] the following are the *principal Components to be considered related to HTA*: Safety and Effectiveness; Economic Impact; and Ethic and Social Impact. The scope refers to an integral approach which is required to understand and to apply HTA's processes adequately.

Related to the different interests present in the health sector's arena, it is clear that the level of quality of the use of information provided by HTA and the results expected from the findings that HTA will provide, both depend on the complexity of the context and on the level of capacity building of the professionals in charge of the process.

The implementation of HTA should be a process planned and supported by policies and strategies focused on the context and needs of the location where the technology is applied. The relevant differences between developed and developing countries demand bringing special attention to identify and

Table 1.2 Decisions of HTA's Stakeholders

Stakeholders	Decisions on
Government	Regulatory approval, reimbursement, public health programs
Health Care professionals	Adoption of technologies, practice guidelines
Hospital and other health care administrators	Equipment procurement, availability of procedures, service delivery
Private sector insurance	Scope and extent of coverage
Manufacturing Industry	Product development, marketing
Patients, health-workers and their representatives	Guidance for treatment and support, access to services; shared decision making with health care professionals
General public, citizens	Information for future decisions on healthcare

Based on International Network of Agencies for Health Technology Assessment—INAHTA. Guidance Document. HTA Agencies and Decision Makers. May 2010. <http://www.inahta.org>; May 2010.

Table 1.3 Stakeholders and Types of Decisions Informed by HTA

Organizations or Individuals	Types of Decisions
Government, agencies, parliaments	Regulatory approval, reimbursement, public health programs, research funding
Health care professionals	Adoption of technologies, practice guidelines
Hospital and other health care administrators	Equipment procurement, availability of procedures, service delivery
Private sector insurance	Scope and extent of coverage
Manufacturing industry	Product development, marketing
Patients, caregivers, and their representatives	Guidance for treatment and support, access to services; shared decision making with health care professionals
General public, citizens	Information for future decisions on health care
Legal professionals	Judges' decisions after demands for the use of high cost health care technologies
Academia	Information for future health care professionals, decisions on research

International Network of Agencies for Health Technology Assessment—INAHTA. Conceptual Paper. The Influence of Health Technology Assessment. <http://www.inahta.org>; April 2014.

understand the priorities and to measure the level of capacities in order to take the best alternatives to apply or even start HTA process. United States, Canada, and United Kingdom provide good examples of advances in Health Sector. In the Latin American Region, Argentina, Colombia, and Mexico have all shown relevant progress in this regard.

Table 1.2 is related to *HTA's Stakeholders* and their respective decisions.

Related to the design of policies, strategies, and the elaboration of plans and programs, one capacity particularly required is having the correct understanding of the value obtained of the implementation of HTA [7]. In the same perspective, the process of identification of priorities and the definition of their consistent solutions using HTA's information is a process which demands a team's effort, sustained exchange of knowledge and experience, and sharing commitment to the defined objectives linked to HTA.

Table 1.3 presents the type of HTA decisions which are expected to improve the level of quality.

1.6 INTERVENTION LEVELS

HTA plays an important role promoting Innovation. It also supports the improvement of the outputs expected as we summarize in Fig. 1.8.

1.6.1 Levels of Influence

HTA provides distinctive support to both quality and safety which are key factors for the health system and specially appreciated by the patients.

The investment of time, budget, human resources, etc., is consistent to the interest on the information provided by HTA's implementation. Let's observe the different levels of influence related to HTA [8]:

1. On policy and administrative decisions: Much of the focus on HTA influence has been in these areas.
2. On subsequent administrative action: Administrative action is dependent on the availability of effective machinery and the willingness of the decision maker to make use of it.
3. On delivery of health care and on the health status of patients: Changes to health care and/or health outcomes related to a health technology may have a weak link to an HTA report. Influence of an HTA report on subsequent action and outcomes within a health care system depends on the actions of many individuals and organizations.

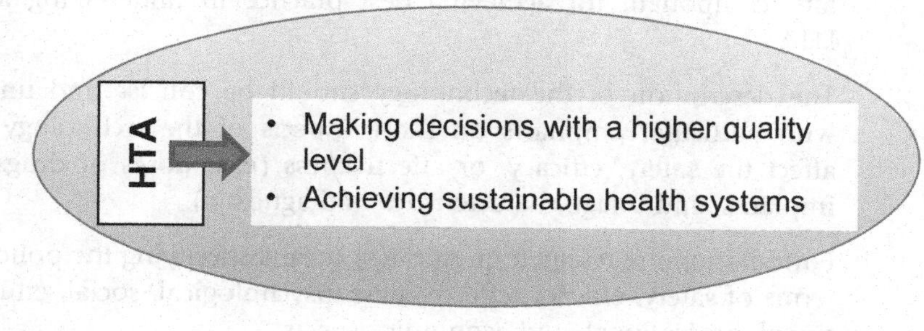

FIGURE 1.8
Health technology assessment's influence improving the outputs, August 2016.

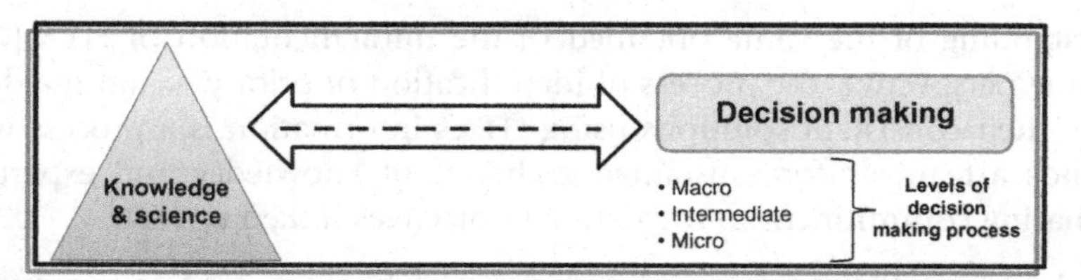

FIGURE 1.9
Health technology assessment's & decision making levels, August 2016.

1.6.2 Levels of Intervention

According to the articulation which is existing in the countries around the world, health systems at the Ministries of Health (MoH) level respond to the Macro or the Central level; related to Intermediate systems the relationship is with Regional or State levels and the Microsystem respond to organizational framework or hospital level (Fig. 1.9).

Related to the several existing gaps in developing countries, it is adequate thinking of a model of a public assessment agency installed and supported by the government including interaction with organizations at intermediate and micro levels, respectively.

1.7 HEALTH TECHNOLOGY ASSESSMENT: PROTOCOL, REPORT

An HTA protocol has to be understood as the elaboration of the plan for both undertaking the whole process of the assessment and writing the HTA report. An HTA protocol has to be understood as the elaboration of the plan for both: undertaking the whole process of the assessment and writing the HTA report. The utilization of such a protocol should be seen as an important component for achieving best practice in undertaking and reporting HTA.

The description of the technology should be concise and understandable, with particular emphasis on those aspects of the technology that directly affect the safety, efficacy, or effectiveness (e.g., doses of drugs, material in implants, and image characteristics of diagnostic).

Formulating the research question(s) means specifying the policy question in terms of safety, efficacy, effectiveness, psychological, social, ethical, organizational, professional, and economic aspects.

The research questions required to be done need to be formulated in an understandable and answerable way, and should be limited in number. Characteristics of research questions include clearly worded, answerable, limited in number, address meaningful outcomes, and address other relevant treatment alternatives.

WHO defines the following trends in HTA (WHO Workshop, Bangkok, September 2010): (a) greater emphasis on cost-effectiveness and economic impacts; (b) rapid reviews/checklists; (c) using surrogate endpoints; (d) using evidence from real-world practice (registries, surveillance, databases); (e) qualitative research (narrative synthesis); (f) tailoring HTA methods to particular types of technology; (g) looking at contexts; (h) international collaboration in HTA methods; (i) reports; (j) including patients views; (k) including needs for training (procedures and devices).

Considering the gaps related to policies, regulation, access to information and alignment, developing countries are closer to the trends (a), (b), (d), (f), (g), and (k).

General Methodology to answering questions:

The following are the principal considerations to be taken in count in the process:

1. Searching for sources of information
2. Selecting and evaluating information (application of inclusion and exclusion criteria)/appraising the evidence
3. Synthesizing the obtained data

1.8 HEALTH TECHNOLOGY MANAGEMENT AND HEALTH TECHNOLOGY ASSESSMENT

The political structure of the country is one of the factors that distinctively affect the HTA process, although in general the trend is having a global approach applied to the HTA process.

Around the world, HTA is provided by government agencies, private institutions, nonprofit organizations, and the academia. In the case of developing countries, it is more adequate to consider the public sector mixed with nonprofit organizations to be the best providers of HTA. In order to strengthen the HTA's progress in the region on 2011, Argentina, Peru, Bolivia, Brazil, Chile, Colombia, Costa Rica, Cuba, Ecuador, México, Paraguay, and Uruguay agreed on the creation of REDETSA as the network to support decision making on incorporation, dissemination, and use of technologies, looking at the contexts of health systems at country level and expanding their access to equity. This initiative was supported by the Panamerican Health Organization—PAHO and Health Technology Assessment International—HTAi.

1.9 HEALTH SYSTEM AND THE IMPLEMENTATION OF HEALTH TECHNOLOGY ASSESSMENT

HTA reports (Three parallel panel sessions at HTAi Workshop, Brazil, 2011) were traditionally strong on assessing the effectiveness and cost-effectiveness of the technology. This is not the case if we observe the implementation HTA on policy decisions, management, organization of care, etc. It is evident that even though HTA is gradually better understood and its lessons and practices are being promoted around the world, the successful implementation of HTA's process is not aligned around the world.

Related to Latin America (PAHO, Argentina, 2013), the gap [9] is represented by (1) the lack of setting explicit links between HTA and the decision of incorporation; (2) the limited number of human resources qualified to lead HTA processes; (3) the growth of the occurrence of indictments in some countries of the Region. The context demands a network between the countries of the region in order to improve the sustainability of the efforts and the results expected by the permanent exchange of information, the discussions of best practices and the joint of efforts to improve capacity building on research and critical appraisal. Two aspects [10] which explain the complexity involved on developing HTA in Latin America are (1) the lack of access to data sources and (2) the weakness of the institutional framework related to the public health sector.

References

[1] NIH Plan for Accelerating Technology Transfer and Commercialization of Federal Research in Support of High Growth Businesses.

[2] Guide 1. How organize a system of healthcare technology management. 'How to manage'. Series for healthcare technology. WHO, Ziken International Consultants UK, PAHO, GTZ, DFID.

[3] CENETEC. Certificado de Necesidad de Equipo Médico. Centro Nacional de Excelencia Tecnológica en Salud. Secretaría de Salud – México; 2017. <http://www.cenetec.salud.gob.mx/contenidos/biomedica/cert_nec.html>; April 13 2017.

[4] European Network for Health Technology Assessment—EUnetHTA. <http://www.eunethta.eu/>.

[5] International Network of Agencies for Health Technology Assessment—INAHTA. Guidance Document. HTA Agencies and Decision Makers. <http://www.inahta.org>; May 2010.

[6] World Health Organization—WHO. Health topics. <http://www.who.int/topics/technology_medical/en/>.

[7] International Network of Agencies for Health Technology Assessment—INAHTA. Conceptual Paper. The influence of health technology assessment. <http://www.inahta.org>; April 2014.

[8] International Network of Agencies for Health Technology Assessment—INAHTA. Working Group 3. Measuring and reporting the impact of health technology assessments. <http://www.inahta.org>; November 2002.

[9] Battista RN, Hodge MJ. The evolving paradigm of health technology assessment: reflections for the millennium. Can Med Assoc J 1999;160(10):1464–7.

[10] Pichon-Riviere A, Augustovski F, Rubinstein A, Martí SG, Sullivan SD, Drummond MF. Health technology assessment for resource allocation decisions: are key principles relevant for Latin America? Int J Technol Assess Health Care 2010;26:421–7.

CHAPTER 2

Health Technology Policy and Regulation

Improving health outcomes and effectively addressing the health needs of the population.

2.1 DEFINITIONS

Health sector technology shows a rapid dynamic of change; on the other side, developing countries gradually bring evidences about the increasing gap in this regard between them and developed countries, the first ones are not able to apply neither to provide an adequate use of health technology.

The situation described influences in the level of quality of diagnostic; therapeutic devices; and procedures corresponding to Macro, Intermediate, and Micro level. In average developing countries have not established and effective national system related to Health Technology Transfer neither a systematic evaluation of new health technologies. According to the World Health Organization—WHO, Health Technology Transfer (Local Production for Access to Medical Products: Developing a Framework to Improve Public Health. Geneva, World Health Organization, 2011. (http://www.who.int/phi/publications/local_production_policy_framework/en/index.html, accessed 10 July 2012).) is the transfer of technical information, tacit know-how, performance skills, technical material or equipment, jointly or as individual elements, with the intent of enabling the technological or manufacturing capacity of the recipients. In the case of medical devices it represents the collaboration of knowledge and resources toward developing medical devices useful for public health needs.

Kwankam [1] states that Health Technology includes devices, drugs, medical, and surgical procedures—and the knowledge associated with these—used in prevention, diagnosis, treatment of diseases, rehabilitation, the organizational and supportive systems required to provide healthcare. Table 2.1 shows the scope of the definition provided.

Healthcare Technology Management Systems. DOI: http://dx.doi.org/10.1016/B978-0-12-811431-5.00002-3
© 2017 Elsevier Inc. All rights reserved.

Table 2.1 Health Technology: Scope to be Considered

Health Technology	Devices, Drugs, Medical and Surgical procedures, Knowledge Organizational/ physical infrastructure Supportive/logistical systems

Health policy is aimed at improving health outcomes; to be effective it should respond to the demand of the population; in this regard and as stated by Banta [2], policy's framework involves

1. vision
2. situation analysis
3. strategy
4. plan of implementation
5. leadership
6. governance

The process described demands a staff in charge for the elaboration of policies within certain skills: an adequate level of knowledge and sustained training in the topics needed; a multi- and transdisciplinary perspective; and the capacity to relate ethics regarding the implementation of policies.

Beyond the level of development of the country's economy, health technology policies should always and without exception address inequity, accessibility, affordability, and availability of innovative as well as core medical devices, responding effectively in all the cases to the health needs of the population; this perspective includes context and culture as relevant factors to develop health technology policies. The understanding and accomplishment to this statement makes sense to provide value and achieve the objectives stated.

Related to medical devices there are four phases to be considered according to WHO [2]: (1) research and innovation; (2) regulation for device safety; (3) assessment for better decision making; and (4) comprehensive management. These phases next to be considered should be adapted to the priority public health conditions, resources, and settings which correspond effectively to the country's population.

2.1.1 Policy

Policy is the declaration of intention to do something, guidance, guide for action based on a set of guiding principles or values, aimed at influencing and defining the decisions and actions of long-term technology for healthcare.

As it is remarked by Walt [3], Health Technology Policy involves actions that affect institutions, organizations, services, and funding arrangements of the healthcare system.

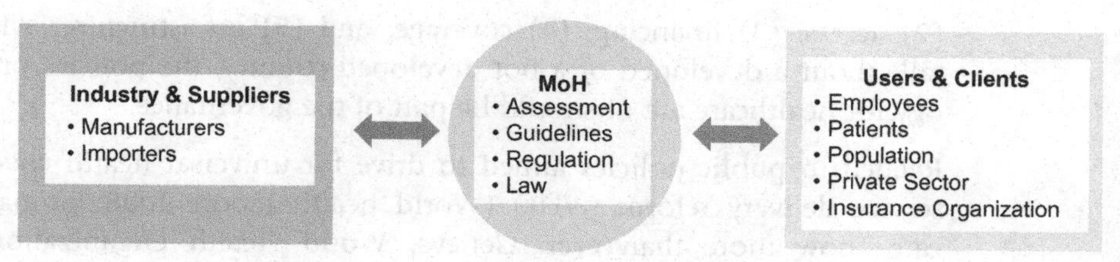

FIGURE 2.1
Health Technology Policy: the stakeholders. *Adaptation of "Healthcare Technology Policy Framework," WHO, 2001.*

WHO states that the effectiveness of the process of formulation policy requires the determination of physical assets, human resources, conducive environment, sociocultural aspects, legal framework, stakeholder participation, governance, donor policies, and public/private sector (Fig. 2.1).

2.1.2 Health Technology Policy and Developing Countries

A relevant challenge for developing countries is the poor effectiveness of the national programs which aims to support the policies; according to WHO in average the initiatives of e-health and telemedicine projects in low resources countries have shown successful results although there are still relevant difficulties to achieve the next step: the design of telemedicine programs based on policies for the country.

In this regard the lack of measurement of the scope and the exact assessment of the results of the national programs over the diverse organizations involved are additional factors to be considered. Interesting and useful information related to the situation described is registered by the Panamerican Health Organization—PAHO, the World Health Organization—WHO, and the Economic Commission for Latin America and the Caribbean—ECLAC.

According to Borowski [4], decision-makers in government prefer to incorporate or balance other factors and information beyond hard evidence when making decisions. Consequently, best practice in policy work requires openness to the kinds of evidence that can or should be used to inform a decision and that the social and political sciences have as much to offer as the natural or physical sciences. This statement brings again the attention to support the process by human resources aligned and trained adequately.

2.2 THE IMPORTANCE OF POLICY FOR HEALTHCARE

National health policies have an important role in the health of the population of a country.

Policies determine the number of key aspects which define the level of response of healthcare system to the population's needs as (1) quality;

(2) access; (3) financing; (4) coverage; and (5) infrastructure. Whether we talk about a developed or a not developed country, the policies on technology for healthcare are or should be part of the governance.

Related to public policies aimed to drive for universal health coverage and service delivery reform. WHO (World health report 2008: primary health care—now more than ever. Geneva, World Health Organization, 2008.) states that effective public policies in the health sector are

- health systems' policies (related to financing, essential drugs, technology, human resources) on which primary care and universal coverage depends;
- public health policies that address priority health problems and include prevention and health promotion;
- policies in other sectors, known as "health in all policies," which call for intersectoral collaboration to achieve positive health outcomes.

According to WHO each of the health systems' policies should not be considered singly but rather with other elements in the system like financing, information, service delivery, health workforce, and governance, see Fig. 2.2.

In developing countries it is recommended to work focusing on the definition of a system of incentives that contribute to make sustainable the proposal of a Change based in Health Technology.

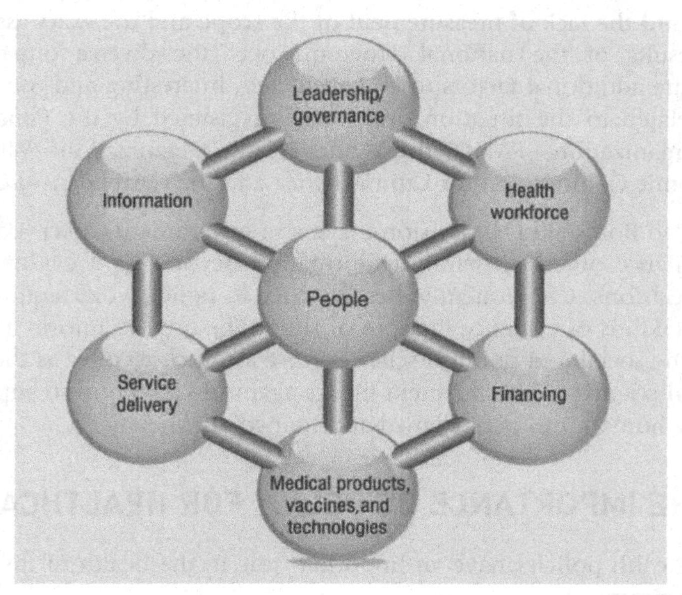

FIGURE 2.2
Six building blocks for health systems, development of medical device policies, WHO 2011.

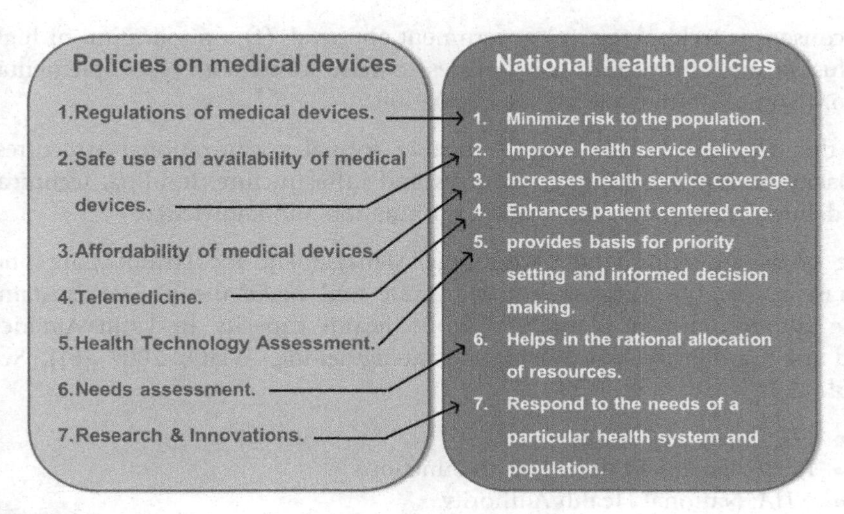

FIGURE 2.3
Relationship between policies on medical devices & national health policies. *Hansen M. WHO global model regulatory framework for medical devices. Geneva: Department of Essential Medicines and Health Products; 2016.*

The process of defining Technology Policy includes promoting (1) the interaction between academia–industry–government, and (2) the definition of governmental mechanisms to raise the visibility of the scientific and technological activity in the countries of the region. See Second Latin American Conference on Research and Innovation for Health, PAHO–WHO Global Forum for Health Research, Panama, 2011.

Observing how the policies on medical devices (health technologies) should focus on matching to national health policy, Hansen [5] provides examples of medical devices (left) and how they match national health policies, see Fig. 2.3:

Some purposes of health technology policies are deeply related to process of accomplishment to the objectives of the national health plan; in this regard the relevance of having the correct understanding, knowledge, and expert staff by the side of the Ministry of Health and other organizations members of the government and related to health sector is evident.

2.3 POLICY PRINCIPLES FOR HEALTH CARE

The World Health Organization (WHO) promotes that policy on Health care technologies includes (a) political aspiration, (b) political will, (c) a complete situational analysis care technology health, (d) the creation of a policy of consensus technology, (e) government approval, (f) a publication of high diffusion and distribution on policies to raise awareness, (g) implementation, (h) monitoring, and (i) the evaluation.

In this regard, policies should include formal organizational structures, capacity building of human resources and infrastructure, funding, technical guidelines, regulations, and sharing information and knowledge.

The goals are improving Technology Management for Health Care and infrastructure; the purpose is to integrate and make the policies sustainable through proper planning (Public health capacity in Latin America and the Caribbean: assessment and strengthening, WHO, 2007 [6]). See Table 2.2.

- *PH*: Policy on Health
- *EPHF*: Essential Public Health Functions
- *NHA*: National Health Authority
- *IHR*: International Health Regulations

General principles to be applied when a new Health Technology Policy is established are (1) a foundation in law; (2) consistency; (3) effectiveness;

Table 2.2 Selected Assessment Tools/Instruments to Assess Policy on Health—PH Capacity

	Selected Tools/Instruments to Assess the Status of PH Capacity				
	Elements of PH Capacity				
Selected Assessment Tools/Instruments	**Workforce**	**Information Systems**	**Health Technologies**	**Institutional and Organizational Capacity**	**Financial Resources**
Tools Developed by PAHO and WHO					
EPHF performance assessment tool	X	X	X	X	
Instrument for performance evaluation of the NHA Steering role				X	
Evaluation instrument of surveillance response capacities (IHR)*		X	X	X	
Guide to rapid assessment of human resources for health (WHO)	X				
Health metrics network framework (WHO)		X			

* forthcoming. *Panamerican Health Organization, Public Health Capacities in Latin America and the Caribbean: Assessment and Strengething, 2007, WHO, Washington DC [6].*

(4) efficiency; (5) impartiality; (6) clarity; (7) transparency; and (8) flexibility. See WHO Global Model Regulatory Framework for five medical devices, 2016.

2.4 HEALTH CARE TECHNOLOGY REGULATION

From the perspective of implementation, at the end of the period of compilation of policies it is required organizational structures aimed to the implementation of strategies, after which the elaboration of action plans is considered in the process. In this regard, Hansen remarks the interaction between the processes of Policy Making and Health Technology Assessment, and this last one is focused on improving the Decision-making domain including regulation. See Fig. 2.4.

Related to the actors, it is needed to define a regulatory authority, also to establish regional and national institutions aimed to assess health technology and of course to manage it [7].

In this phase observe that some key aspects are related to (1) have specialized and certified professionals in biomedical engineering and other related areas such as Health Technology Assessment, Clinical Engineering, and

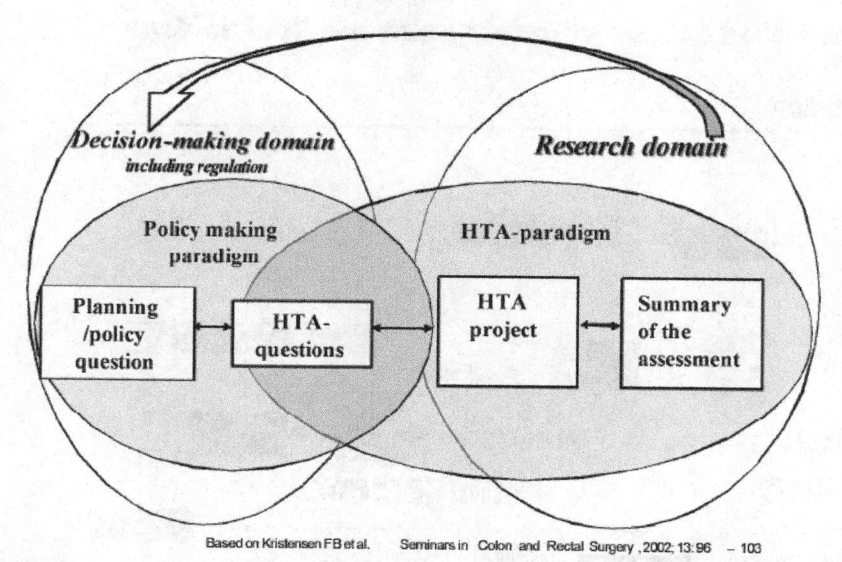

FIGURE 2.4

Policy making and health technology assessment. *Hansen M. WHO global model regulatory framework for medical devices. Geneva: Department of Essential Medicines and Health Products; 2016.*

Health Technology Planning; and (2) the design and the utilization of indicators to measure and control the effectiveness of the policies.

2.5 HEALTHCARE POLICY: GENERAL OBJECTIVES AND SCOPE

Policy-makers, decision-makers, health staff of units whose work influence health objectives (Center for Medical Technology Policy is another reference http://www.cmtpnet.org/.) are engaged to the level of quality of the implementation of health policies, related to this point, see Fig. 2.5. In addition, an adequate framework to develop policies for health technology is required. Regarding the framework, the quality of the policy is defined by the correct understanding and knowledge of the culture and the context.

Some factors that influence the effectiveness of healthcare policies are as follows:

1. Health policy should be aimed and support the development of health technology management staff; they should have career possibilities and also adequate working conditions. In this regard, the organization will define mechanisms and human resources strategies to improve the sustainability of the staff.

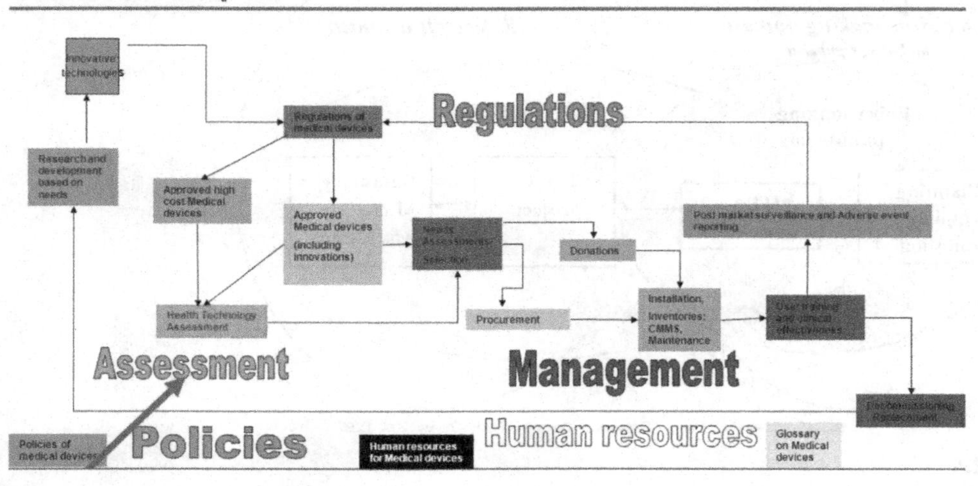

FIGURE 2.5
National health policy framework, Development of Medical Device Policies, WHO, 2011.

2. Knowledge and correct approach to the culture and context of the country to improve the policy alignment.
3. Budgeting available to cover permanent training and continuing education to support the growth and competitiveness of the staff.
4. Investment on plans to provide advocacy activities that promote (i) the establishment and application of norms and standards in the utilization of health technology; (ii) the role of health technology in health services by exchanging with politicians, health planners, decision-makers, and health staff.

References

[1] Kwankam Y. Health Care technology framework. Alexandria, Egypt, WHO; 2001.

[2] Banta H. Assessing medical technologies. Washington DC: Institute of Medicine, National Academic Press; 1985.

[3] Walt G. An introduction to health policy-process and power. Chicago, US, Oxford University Press; 1994.

[4] Borowski H. Linking evidence from health technology assessments to policy and decision making: The Alberta Model. Int J Technol Assess Health Care 2007.

[5] Hansen M. WHO global model regulatory framework for medical devices. Geneva, Switzerland, Department of Essential Medicines and Health Products; 2016.

[6] Panamerican Health Organization, Public health capacities in Latin America and the Caribbean: assessment and strengthening, 2007, WHO, Washington DC.

[7] Krech L. Health capacity in Latin America and the Caribbean: assessment and strengthening. Washington DC, US, Panamerican Health Organization—PAHO; 2007.

CHAPTER 3

Human Resources and Healthcare Technology Workforce

Healthcare Technology Management's staff responding consistently and with effectiveness.

3.1 ASPECTS ABOUT HEALTH AND HEALTH PROFESSIONALS

The first aspect is *"Health is all about people"* as it is stated by Frenk [1]; who is clear in observing the interaction between the set of people requiring health services and the workforce in charge of health service's delivery.

A second aspect according to Frenk is *Trust*, a key factor which influences the user's decision related to the selection of a determined health service from a number of other ones, he states that Trust is a mix of technical competence and service orientation, additionally ethical commitment and social accountability complete the definition of health professional work. The value of trust is distinctively present and demanded in healthcare system; the process is independent of the personal budget available by the user of health services. Fig. 3.1 illustrates the process according to Heckley [2].

Heckley drives us to observe and to analyze correctly the dimension of influence of Trust in health sector, consistently to this perspective the users expect always from healthcare organizations a positive and aligned feedback responding to the trust he provides among his interaction with health system.

Health workforce is the third aspect to be considered, and it is the coordinated effort and teamwork of health workers at the organization. Health's goals nowadays demand necessarily to plan and to act as a health workforce. In this regard, Frenk analyzes the strong role of health staff as human linkage connecting knowledge to action. The process described certainly depends on the culture, values, economic context, and policies of the country, consequently is valuable to invest budget, time, and other resources to promote and improve the organization's capacities to develop workforce's culture.

Healthcare Technology Management Systems. DOI: http://dx.doi.org/10.1016/B978-0-12-811431-5.00003-5
© 2017 Elsevier Inc. All rights reserved.

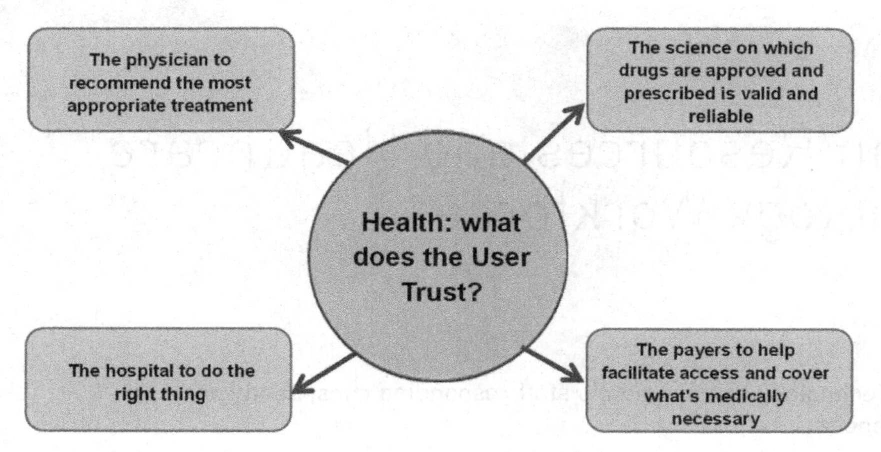

FIGURE 3.1
HealthCare and the value of trust. *Based on Heckley P. The healthcare workforce and public trust: what they think matters, 2016.*

Aspect number four is related to the fact that *Health sector in developed and developing countries is incorporating gradually more technology*: "soft": processes, systems, strategies, etc., and "hard": devices, equipment, etc., the process states a continuous change and drives to the organization to be more complex and costly. In addition and consistently with the country's context and needs, the role of human resources is key to improve health services, achieving health goals and making the best decisions related to health technology in benefit of the patient. Related to this point, statistics show health systems have difficulties to provide health staff on time around the world ("Global strategy on human resources for health: workforce 2030," WHO, 2016.).

The fifth aspect regards the *Three Generations of Health Educational Reforms* occurred among the past century according to Frenk:

1. The first generation, at the beginning of the 20th century, taught a science-based curriculum.
2. Around the mid-century, the second generation introduced problem-based instructional innovations.
3. A third generation is systems based to improve the performance of health systems by adapting core professional competencies to specific contexts incorporating global knowledge. See Fig. 3.2.

Finally *Equity* as the sixth aspect [3] demands to mobilize leadership within the educational and health systems, to invest more, to develop quality control mechanisms, and to strengthen global learning. This perspective is not related to setting kind marketing messages; but is strongly engaged with

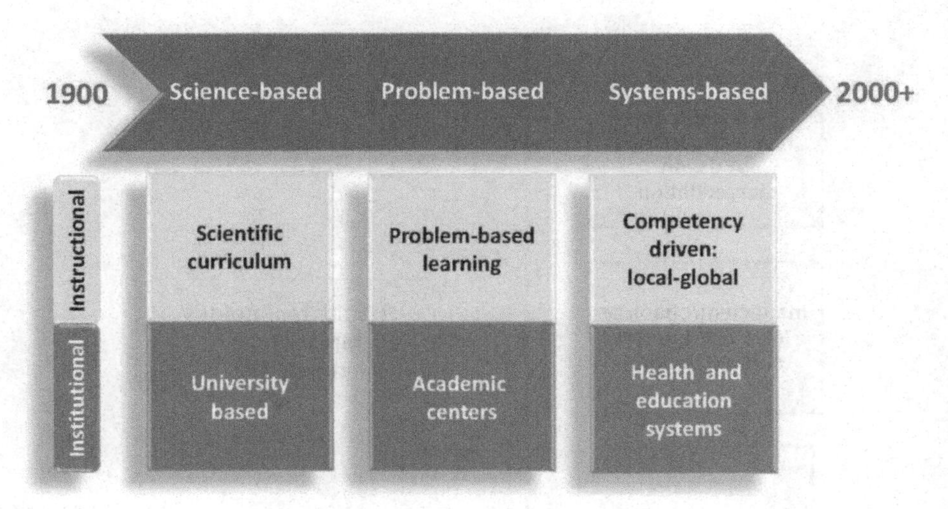

FIGURE 3.2

Health educational reforms. *Frenk J, Chen L, Bhutta ZA, Cohen J, Crisp N, Evans T, et al. Health professionals for a new century: transforming education to strengthen health systems in an interdependent world. The Lancet 2010;376(9756):1923–58.*

processes, strategies, and policies to make Equity visible through the outputs of the health organization.

3.1.1 Health Organizations: Complexities and Challenges

Over the last 20 years health sector is characterized by (1) the growth of knowledge and technologies; (2) health staff are expected to be able to successfully deal with complex care management, expanding functions, and different types of facilities: home-based, community-based, school-based health centers, etc. (U.S. National Library of Medicine, Department of Health and Human Services, National Institutes of Health, 2016.).

The complexity of the context described above is especially challenging for developing countries (Fostering Health Technology Management, Biomedical Engineering & Innovation Competitiveness: Health Technopole CENGETS in Peru, Rivas R., Vilcahuamán L., 2016.), related to health technology some of the principal obstacles human resources deal with at developing countries are

1. lack of human resources trained and with accreditation;
2. insufficient/obsolete/inexistent norms and regulations;
3. lack of incentives as Fig. 3.3 illustrates.

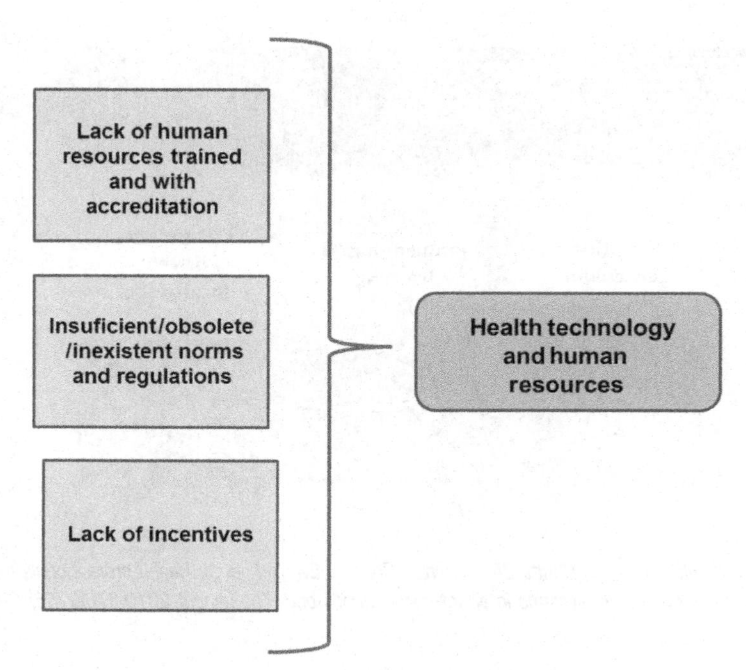

FIGURE 3.3
Obstacles for health staff related to health technology. *Created by Rivas (2016).*

The following are considerations related to health staff's challenges (Frenk, 2010): (1) health professionals should acquire competencies matching needs; (2) competencies should be connected globally; (3) a culture of critical inquiry and the effective use of information technologies should be included.

Health's trends and goals demand to change the traditional concept related to hospitals role and health staff's profile, the achievement of high-value results depends on a number of factors: one of them is the understanding of the science–technology–ethics link; this synergy is part of the core of the level of Quality and Efficiency expected from Health systems.

In this regard Ribera [4] provides insights as a result of an interesting research applied in two hospitals, see Table 3.1.

Hospital's goals demand to promote the development and improvement of a sustained relationship between the healthcare organization and the academia: biomedical engineers, clinical engineers, physics, health technology managers, etc., oriented to create a collaborative partnership focused in

Table 3.1 Challenges of Leading Public Hospital

Challenges of Leading Public Hospital in the new Context	Participation in the detection, adoption, and dissemination of healthcare service innovations in specialized services
	Inclusion of Clinical and Patient-Flow Services in Healthcare process innovation
	Coordination with stakeholders to improve management and efficiency of health services

Ribera J, et al. Hospital of the future: a new role of leading hospitals in Europe. Barcelona: IESE Center for Research in Healthcare Innovation Management; 2016.

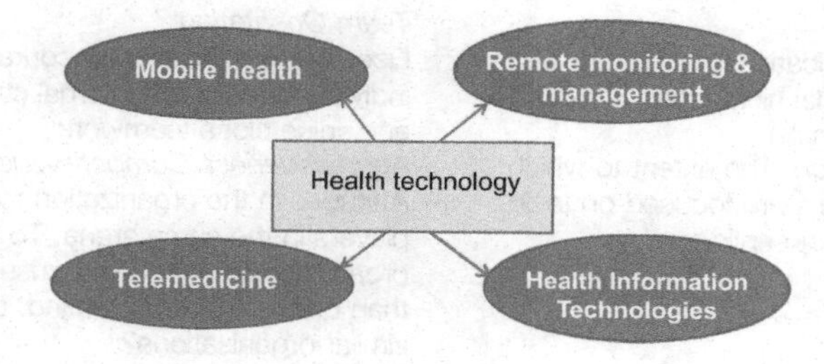

FIGURE 3.4
Health Technology: The improvement of health services, elaborated by Rossana Rivas, 2017.

health technology projects which matches effectively the needs of the health organization and their users.

The level of quality expected from health service's users' needs an interdisciplinary and multidisciplinary teamwork. It is expected in this regard; the members design and implement improvements focusing on high impact and cost-effective's projects. In this perspective, health technology has gradually a more relevant role in the process of improvement health services as it is shown in Fig. 3.4.

3.2 ORGANIZATIONAL CULTURE'S: RELEVANCE AND INFLUENCE ON HEALTH STAFF'S WORK AND OUTPUTS

Organizational culture plays a distinctive role in the processes of development and improvement of competitiveness and performance of health organizations.

Despite there are a number of differences between the countries around the world all the organizations and specially health's organizations are influenced by the different aspects related to organizational culture. In this regard,

Table 3.2 Aspects of the Organizational Culture

Aspects of Organizational Culture: Relationship with Health Organization's Performance & Competitiveness	
Attitudes to Innovation and Risk Taking: Whether the organization encourages and rewards new ways of doing things, or instead values and maintains traditional approaches	*Uniformity or Diversity*: The attitudes and expectations within the organization that either value consistency or encourage diversity
Degree of Central Direction: The extent to which objectives and performance expectations are set centrally rather than being devolved	*People Orientation*: This dimension encapsulates the attitudes toward, support for, and valuations of, the organization's human resources
Patterns of Communication: The degree to which communication, instruction, and reporting are restricted to formal hierarchies of authority (compared with informal channels)	*Team Orientation*: Does the organization encourage and reward individualism, or are internal structures designed to foster and value close teamwork?
Outcome or Process Orientation: The extent to which control and reward mechanisms are focused on tasks compared with the end product/service	*Aggressiveness/Competitiveness*: Attitudes in the organization toward other external players in the same arena. To what extent are organizational attitudes focused on dominating rather than coexisting, cooperating, or even learning from other similar organizations?
Internal or External Focus: The extent to which attention is directed at external stakeholders, especially customers and the wider community, compared with an emphasis on internal organizational issues	*Attitudes to Change*: To what extent is the organization focused on internal stability rather than dynamic concerns (such as increasing size, scope, or competitiveness)?

Adaptation of Davies based on Robbins SP. Organizational behavior: concepts, controversies and applications, 7th ed. Englewood Cliffs: Prentice Hall, 1996 [5]; Newman J. Shaping organisational cultures in local government. London: Pitman, 1996 [6].

to know and to understand these aspects as well as to be able to manage them, we will contribute to achieve the objectives expected.

Table 3.2 illustrates some of the aspects which are part of the organizational culture according to Davies [7].

To identify particular cultural attributes that are facilitative of performance is not simple although to do it is highly required from healthcare organizations due to the complexity of health sector and the direct and permanent contact of the health staff with the patients.

Davies remarks the organization which intends to change the culture within healthcare will need to heed the constraints imposed by external influences on cultural values. This perspective has sense thinking about the diversity of healthcare professions inside of the organization and the different levels of use of technology they apply.

The achievement of goals requires from the health organization to be able to implement an effective interaction with the external stakeholders, Daft [8] contributes to understand the diversity existing in the organizational environment; he divides it into 10 sectors:

1. Industry sector: related businesses and competitors
2. Raw materials sector: suppliers, manufacturers, service providers
3. Human resources sector: employees, labor unions, schools, colleges, employment agencies, labor markets
4. Financial resources sector: banks, lenders, stock markets, investors
5. Market sector: actual and potential customers, clients, and users of products and services along with their characteristics and preferences
6. Technology sector: science, technological methods of producing products and services
7. Economic conditions sector: levels and rates of employment, inflation, growth, investment, and other economic circumstances
8. Government sector: laws, regulations, court rulings, political systems, and governmental services at the local, state, and federal levels
9. Sociocultural sector: characteristics of society and culture such as age, education, values, and attitudes
10. International sector: other countries and globalization

Finally the correct understanding and analysis of the external information requires to be informed about and to act consistently to the trends of healthcare, see Davies insights in this regard:

I. Patients becoming more informed consumers
II. Growth of structured quality measures
III. Revenue-driving consolidation
IV. New and alternative provider payment models
V. Specialty drug use driving the cost of care
VI. Information technology innovations driving inter-stakeholder communications

Related to the Organizational factors that impact on the health worker performance, Lundstrom [9] emphasizes on (1) Delivery Systems; (2) Specific practices; (3) Teamwork, errors, attitude, and stress: Impact on performance; (4) Influence of quality improvement on worker satisfaction.

3.3 HEALTH CHALLENGES AND COMPETENCIES OF HUMAN RESOURCES

3.3.1 Complex Challenges

According to Frenk, scientific knowledge produces new technologies and empowers citizens to adopt healthy lifestyles, improve care-seeking behavior, and become proactive citizens. Knowledge translated into evidence can also guide practice and policy. The author remarks that health systems are socially driven differentiated institutions with the primary intent to improve health and over the last century health professional's contribution have been remarkable, saying this 21st century requires new educational strategies and health staff's (service providers, physicians, caregivers, engineers, managers, communicators, educators, policy makers, etc.) competencies.

Aligned to Mesko, Magennis and Wentzel (Health Information Systems—The Impact of Disruptive Technologies, R. Magennis, H. Wentzel, 2015.) conclude that the "space" between doctor and patient is filling with technologies that will improve preventive, diagnostic, and therapeutic, follow up and outcome medical care.

Health system's core, goals, and challenges are deeply related to technology; around the world health professionals understand the relevance of being able to manage and to make decisions according to this trend, also to use health technology consistently to the context, needs, culture of their respective country.

3.3.2 Competencies

The following are core competencies required for health professionals regardless of their discipline, to meet the needs of the 21st-century healthcare system, as the Institute of Medicine—IOM [10] remarks, see also Fig. 3.5:

1. *Delivering patient-centered care*: identify, respect, and care about patients' differences, values, preferences, and expressed needs; relieve pain and

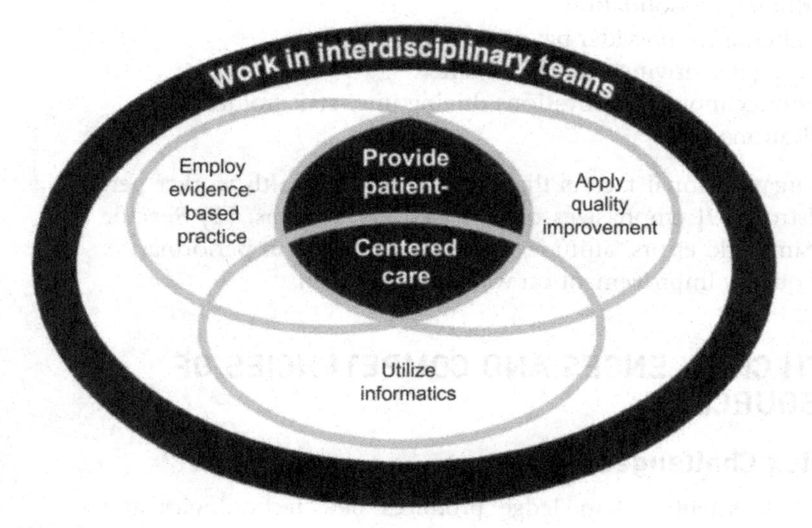

FIGURE 3.5
Overlap of core competencies for health professionals. *IOM. The core competences needed for healthcare professionals, 2003.*

suffering; coordinate continuous care; listen to, clearly inform, communicate with, and educate patients; share decision making and management; and continuously advocate disease prevention, wellness, and promotion of healthy lifestyles, including a focus on population health.
2. *Working as a part of inter-professional teams*: cooperate, collaborate, communicate, and integrate care in teams to ensure that care is continuous and reliable.
3. *Practicing evidence-based medicine*: integrate best research with clinical expertise and patient values for optimum care, and participate in learning and research activities to the extent feasible.
4. *Focusing on quality improvement*: identify errors and hazards in care; understand and implement basic safety design principles, such as standardization and simplification; continually understand and measure quality of care in terms of structure, process, and outcomes in relation to patient and community needs; design and test interventions to change processes and systems of care, with the objective of improving quality.
5. *Using information technology*: communicate, manage knowledge, mitigate error, and support decision making using information technology.

3.4 HEALTHCARE TECHNOLOGY MANAGEMENT: QUALITY & EFFICIENCY RELATED TO HUMAN RESOURCES

Health and Public Policy Committee and Office of Health Policy—HPPC of the Royal College of Physicians and Surgeons of Canada [11] emphasizes on the following eight measurable dimensions of quality aimed to quality improvement in healthcare:

1. Safety
2. Accessibility
3. Acceptability
4. Appropriateness
5. Provider competence
6. Efficiency
7. Effectiveness
8. Outcomes

HPPC provides definitions to support the promotion and the organizational action aligned to the objectives of quality in healthcare, see Table 3.3.

Table 3.3 Dimensions of Quality

Safety: Reduction of "patient safety incidents (adverse events)" within the healthcare system through the use of leading practices shown to improve patient outcomes and enhance prevention.	*Accessibility*: Availability of care, based on medical need, within the healthcare system. Accessibility should be fair, equitable, timely, easy, and affordable. It includes providers, drugs, technologies, facilities, information, redress, and treatments throughout the continuum of care.
Acceptability: Meeting appropriate patient and societal needs by recognizing the informed references of patients and society regarding accessibility, alternative treatments, patient–practitioner relationship, amenities, the effects of care, and the cost of care.	*Appropriateness*: Necessary and ethical care for patients by providers who balance the "primacy of patient welfare" and cultural and social expectations against the management of finite resources.
Provider Competence: Enables the measure of performance and is characterized by knowledge, traits, skills, abilities, and behaviors resulting in quality outcomes.	*Efficiency*: Optimal use of minimal or scarce resources to achieve desired results.
Effectiveness: Degree to which desired health outcomes are achieved with the application of active therapies and treatments.	*Outcomes*: Results measured against objectives, standards, or expectations from patients or providers.

Royal College of Physicians and Surgeons of Canada. The art and science of high-quality health care: ten principles that fuel quality improvement. Health and Public Policy Committee and Office of Health Policy;Quebec, Canada, 2002.

Healthcare Technology engages the following stakeholders of the health organization:

- Clinical staff, physicians, health staff
- High- and medium-level decision makers
- Managers, Directors, Chiefs, Supervisors
- Caregivers
- Patients
- In-house technical staff

Health Technology Management (HTM) is essential for health organization's accomplishment of quality healthcare; this is pertinent for developed and developing economies. Healthcare technology that is out of order quickly leads to a decline in demand, which will in turn reduce the income and quality of services of the health facilities as Lenel [12] states.

HTM provides health and economic benefits for the organization and provides quality health services by meeting health service standards, some examples below are provided by Lenel:

A. Ensuring the safety of technology
B. Developing operational skills
C. Disposing of technology without harming the environment

- Progress with establishing framework conditions and an HTM Unit.
- Administrative and Technological requirements for HTM Unit.
- Cost-Efficiency.
- Working condition of healthcare technology.
- User's satisfaction.
- Sustainability.

FIGURE 3.6
Overlap of core competences for health professionals. *IOM. The core competences needed for healthcare professionals; 2003.*

D. Planning acquisitions that are cost-effective
E. Running services economically

Fig. 3.6 shows the main areas to measure HTM Unit's outputs at the health organization stated by Lenel.

3.5 HEALTHCARE TECHNOLOGY MANAGEMENT AND HUMAN RESOURCES: SOME APPROACHES

Human Factors Professionals and Limitations for HTM are presented here to promote quality, safety, and sustainability related to health technology.

Applying human factors methods and principles can help in identifying where and when issues associated with a mismatch between users, technologies, and environments are likely to occur as Vicente [13] stated. Fig. 3.7 illustrates the interface between human factors and technology a valuable contribution of the author.

Human Factors applied to Healthcare are oriented to provide reliable recommendations to mitigate risks associated with the use of healthcare technology as it is recognized by the Ministry of Health of Canada, the FDA in United States and others.

Cassano-Piché and Easty [14] contribute to understand the value of considering health factors from a healthcare technology perspective and remark the following three factors to explain the failures in safety at healthcare organizations: (1) the number of healthcare technologies present in patient care areas, (2) the complexity of healthcare technologies, and (3) the pace of technological change.

In the other side Table 3.4 shows limitations for HTM's productivity adapted from Weier [15].

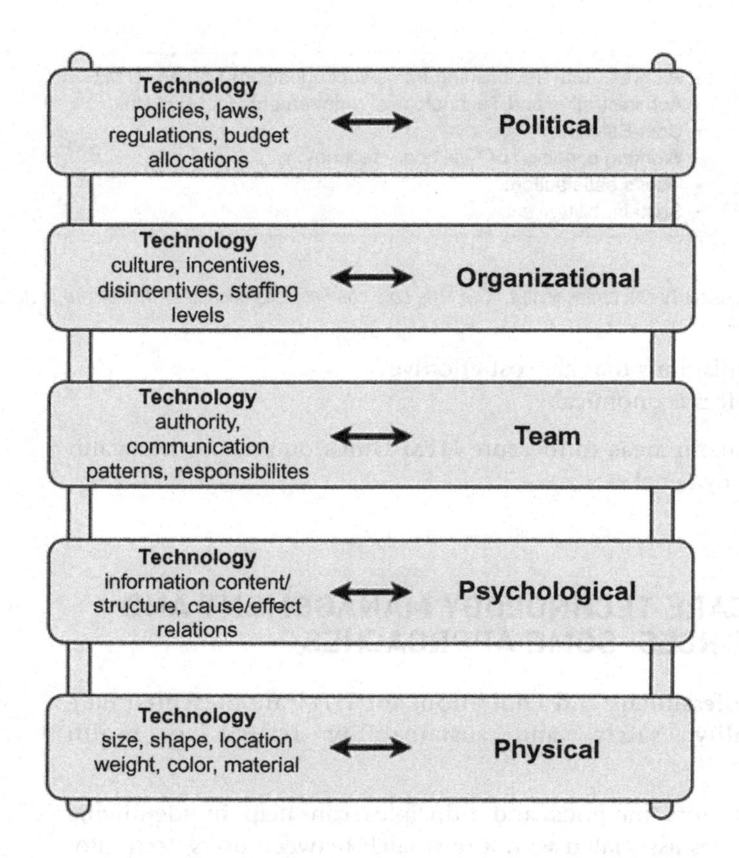

FIGURE 3.7
"Human-Tech Ladder," adapted by Canadian Patient Safety Institute's—CPSI and Patient Safety Education Program—PSEP. *From Vicente (2015).*

Table 3.4 Limitation for HTM's Productivity

Limitations for Health Technology Management's productivity	Lack of political will and long-term commitment throughout the health organization related to the role of technology to achieve health goals
	Perceived lack of time and talent to implement HTM's productivity initiatives
	Perception by health stakeholders, inside and outside the organization, that healthcare technology is not relevant or is a subject which engages only technicians
	Lack or inefficient management of information

Adapted by Rivas from Weier S. Best practices approaches to healthcare labor productivity management. AMN Healthcare; 2016.

Key aspects to improve productivity stated by Weier are also present in the case of health organizations. In this regard the contribution of HTM to the productivity of the organization related to cost reduction, health services restructuring, and clinical restructuring is valuable; a summary of health organization's multiple fronts remarked by Weier is given below:

1. *Cost reduction*: Flexible budgets and budget controls measurement; Expanded productivity targets and reporting.
2. *Business restructuring*: Delivery system rationalization and restructuring; Organizational structure portfolio evaluation and management.
3. *Clinical restructuring*: Care processes redesign and restructuring; Physician Engagement and Integration; Expanded relationships with other providers.

References

[1] Frenk J, Chen L, Bhutta ZA, Cohen J, Crisp N, Evans T, et al. Health professionals for a new century: transforming education to strengthen health systems in an interdependent world. The Lancet 2010;376(9756):1923–58.

[2] Heckley P. The healthcare workforce and public trust: what they think matters, 2016.

[3] World Health Organization—WHO. Transforming and scaling up health professionals' education and training, 2013.

[4] Ribera J, et al. Hospital of the future: a new role of leading hospitals in Europe. Barcelona: IESE Center for Research in Healthcare Innovation Management; Barcelona, Spain, 2016.

[5] Robbins SP. Organizational behavior: concepts, controversies and applications. 7th ed. Englewood Cliffs, NJ: Prentice Hall; 1996.

[6] Newman J. Shaping organisational cultures in local government. London, UK: Pitman, 1996.

[7] Davies H, et al. Organizational culture and quality of health care. Quality in Health Care; York, UK, 2000.

[8] Daft R. Organization theory and design. Nelson Education; 2012.

[9] Lundstrom T, et al. Organizational and environmental factors that affect worker health and safety and patient outcomes. Am J Infect Control 2002;30:93–106.

[10] IOM. The core competences needed for healthcare professionals, Washington DC, US, 2003.

[11] Royal College of Physicians and Surgeons of Canada. The art and science of high-quality health care: ten principles that fuel quality improvement. Health and Public Policy Committee and Office of Health Policy; Quebec, Canada, 2002.

[12] Lenel A, et al. How to organize a system of healthcare technology management, series for healthcare technology. WHO; 2005.

[13] Vicente K. The human factor: revolutionizing the way people live with technology, Toronto, Canada, 2003.

[14] Cassano-Piché A, Easty T, et al. Evaluating and improving the use of health technology in the real world. Human Era, UHN, IFMBE-CED; Toronto, Canada, 2015.

[15] Weier S. Best practices approaches to healthcare labor productivity management. AMN Healthcare; 2016.

CHAPTER 4

Healthcare Technology Planning and Acquisition

Planning technology from design to disposal is as useful as checking the map before traveling.

It is known that the technological environment of the hospital is complex due to its diversity and countless considerations required to guarantee its functionality. Planning is the first process in an integrated healthcare technology management (HTM) system. Even if planning is not absolutely accomplished, it is an important way to deal with the complexity of the technology used in hospitals. On planning, there is a wealth of information on how to propose processes, indicators, planning products, planning tools, etc. We here only want to point out some aspects of context and details of what is characteristic of technology in health, which, in our opinion, leads to the need to rethink the organizational models of hospitals. It is known that a plan is a specific design to achieve a desired future. In the case of a hospital this future is described in all institutional documents and premises, institutional mission and vision, institutional strategic plan, master plan of the institution as a whole, the annual operational plan, and the annual plan of acquisitions. From these institutional plans, each unit usually develops its own planning documents, more specific such as the master plan of a particular clinical service or the annual procurement plan of this service.

What we observe in this context is a diversity of gaps regarding healthcare technology. In some cases the aspects of technology are not included in the planning documents and in other cases an interpretation of the institutional premises is required to identify them and thus generate precise actions on technology management. Usually the vision and mission that the hospitals raise are the quality of health services, they propose to improve the medical specialization and how it will be given, but on technology there are no explicit proposals. The hospital also works to humanize the care of services,

Healthcare Technology Management Systems. DOI: http://dx.doi.org/10.1016/B978-0-12-811431-5.00004-7
© 2017 Elsevier Inc. All rights reserved.

is concerned with the profiles of health professionals and in the topic of technology it is assumed that the maintenance of medical equipment is sufficient to provide the quality of the service proposed. Then we can ask some questions:

- Who is responsible for the technology in the hospital?
- Is the head of the medical department or the head of general services who plans the technology?
- Which unit does set policies on technology?
- Who is responsible for the design of clinical services?
- Even when third parties are hired, a local capacity for planning is required, who is responsible for healthcare technology assessment HTA?
- Who is in charge of determining the clinical effectiveness and then deducing the correspondence with the technology used?
- Who does perform an efficiency analysis of the technologies used and plans the goals of improvement?
- Who does perform specific analysis about resources waste due to the use of technology?
- Who does verify the application of technical and safety standards and sets the goals to be met?
- Finally, what hospital unit does have in its business plan the task of making the technology (as defined in Chapter 1: Healthcare Technology Management (HTM) & Healthcare Technology Assessment (HTA)) functional throughout the hospital?

These and many other details of the technology should fit more appropriately into an organizational structure in order to serve them and mainly to have responsible ones in charge. However, the vast majority of publications and books on clinical engineering and hospital engineering refer only to medical equipment and hospital equipment, respectively, leaving aside the organizational structure that will make them practicable. And it is with this look that we are going to sketch the way of how the planning would be carried out in the framework of an organizational model in charge of the technology, this means an organization with three branches: the medical branch, the administrative branch, and the branch of the Technology, whose main advantage is that it allows to address all these technological issues in a more convenient way through processes, with identifiable leaders, and naturally in coordination with the other medical and administrative branches. It should be noted that in our concept, any successful health intervention entails the need to consider these three aspects: the medical, the administrative, and the technological; lack or weakness of any of these puts at risk the intervention and therefore prevents the fulfillment of the goals raised in health.

4.1 PLANNING THE FUNCTIONAL TECHNOLOGY IN THE CLINICAL SERVICES

The mission of a unit in charge of technology in health (technology branch) may well be posed in these terms: Provide the appropriate technology for the delivery of health services with effectiveness, efficiency, safety, quality, and reasonable costs. We understand technology as indicated in Chapter 1, Healthcare Technology Management (HTM) & Healthcare Technology Assessment (HTA), i.e., we have technologies for individual health services (clinical technologies: clinical procedures, medical devices, drugs, and medical materials; and supporting technologies: organization, information systems, infrastructure, and hospital equipment); and technologies for community health services (protection, prevention, and promotion technologies). The challenge is to plan each of these aspects that make hospitals and their clinical services functional from the technology focus. Who is currently doing it? Health professionals have as their main function the care of health services; administrative staff handles the accounts, the logistics, the payments; Medical equipment, electromechanical equipment, computers, and others are handled as part of the administrative branch, in fact many engineering departments belong to general services. In these circumstances, where the engineering assignment is focused on the medical and hospital equipment, it is very difficult to propose a correct management so that every aspect of the technology is functional (see Chapter 9: New Organizational Model for Hospitals in the New Technology Context).

In order to have functional technology in the hospital and in clinical services, one must first consider leadership in technology, identify or create the unit in charge, and have those responsible, professionals with a clinical engineering profile, biomedical engineering, medical physics, etc., including health professionals themselves. Leadership could be considered as a health sector policy and be a strategy in considering the results and relevant impacts that can be had for example in the increase of the quality of health services or greater coverage of services. Second, this unit or branch of technology must have the function and the task of formulating policies and strategies for the development of technology in the hospital. This function requires having the appropriate human resources and sufficient financial and material resources, as well as having institutional alliances in health technology. In the following chapters we will have more details on these aspects.

Planning, in order to have the appropriate functionality of the hospital's technological environment, requires a holistic view in order to consider all the technological components required for clinical work. Health professionals so require it to do their job. The most developed in technology planning is associated on the one hand to the design of new or renovated facilities and on the other hand to planning of medical devices. Many authors have suggested that clinical engineers should work much more closely with architects and design engineers [1]. However, since these working groups belong to different units, we have a breakup and lose coherence in the technological environment. For example, the ventilation system may be working well but if medical equipment does not work properly, health professionals are no longer in an appropriate technological environment.

The following table shows the planning components according to the type of health technology, in which, in one way or another, the clinical engineers should be involved, knowing that this management involves the intervention of a multidisciplinary team, but not for this the responsibilities must be diffuse. Table 4.1 shows the need for different professional profiles to elaborate the planning components for each type of technology, and in addition, in order to have a functional and operational technological environment, an organic management of the whole is required. It makes relevant to have a technological branch with an integrated HTM system.

Each of the planning components will be dealt with in the following subchapters. Table 4.1 illustrates the complexity of assuring the functionality of the technological environment in a hospital or a clinical service, as well as showing that managing types of technology in an individual way does not always lead to an appropriate technological environment, since clinical work requires full operation of all technology types at the same time.

To reduce this inherent complexity of planning it is better to consider planning for each clinical service and then at a higher level, consider planning the hospital as a whole. This approach tends to the specialization of the clinical or biomedical engineers in relation to the medical specialties, which is a very good practice in the hospital, since it allows a better integration of the professional engineers and the medical assistance personnel. Also, Table 4.1 refers to the planning for the technology to be acquired or incorporated, in another side it is possible to indicate that in the following subprocesses of the technology application cycle in hospitals, each one requires planning of their activities and products, for example in the process of property management can be set goals of operating equipment percentage, reduction of average response time, costs for maintenance and other similar ones, and in risk management processes can be set goals as the number of adverse events recorded, number of hazards identification & risk assessment HIRA matrices, etc. (See more details in Chapter 5: Asset & Risk Management Related to Healthcare Technology).

The products of the planning may be diverse; Table 4.2 shows examples of these products taking into consideration a clinical service and the main processes of an integrated health technology management system. The work of

Table 4.1 Planning Components According to Type of Health Technology

Technologies for Healthcare	Term Planning	Components of Technology Planning				
Individual Health Services						
a. Clinical technologies						
Clinical procedures	Medium term	Needs analysis according to clinical requirements	Technology assessment to options identification	New or renovated facility requirement evaluation	Financial requirement evaluation	Priority setting into planning period
Medical devices	Medium term					
Pharmaceuticals & others consumable products	Short term			Application and impact evaluation		
b. Support technologies						
Organization & management system	Medium term	Needs analysis according to clinical requirements	Technology assessment to options identification	Application and impact evaluation	Financial requirement evaluation	Priority setting into planning period
Infrastructure & hospital equipment	Long term			Master plan elaboration	Financial requirement evaluation & Predesign	Schematic design & Constructive details sheets
Information & communication technologies	Medium term			New or renovated facility requirement evaluation	Financial requirement evaluation	Priority setting into planning period
Community Health Services						
Protection technologies	Short term	Needs analysis according to clinical requirements	Technology assessment to options identification	Application and impact evaluation	Financial requirement evaluation	Priority setting into planning period
Prevention technologies	Short term					
Promotion technologies	Medium term					
Environmental Health	Medium term					

Table 4.2 Examples of Planning Products in HTM System According to Type of Health Technology and Clinical Service

Technologies for Healthcare	Clinical Service or All Hospital					
	Planning	**Acquisition**	**Asset Management**	**Risk Management**	**Human Resources Development**	**Research & Projects**
Individual Health Services						
a. Clinical technologies						
Clinical procedures	Incorporation plan	Incorporation process plan	Update plan	Clinical effectiveness evaluation	Training plan	R&D plan
Medical devices	Annual plan	Acquisition process plan	Maintenance plan	Standards to be met, HIRA plan	Training plan	R&D plan
Pharmaceuticals & others consumable products	Annual plan	Acquisition process plan	Storage & distribution	Standards to be met plan	Training plan	R&D plan
b. Support technologies						
Organization & management system	Incorporation plan	Incorporation process plan	Update plan	Efficiency evaluation plan	Training plan	R&D plan
Infrastructure & hospital equipment	Master plan	Hiring process plan	Maintenance plan	Standards to be met, HIRA plan	Training plan	R&D plan
Information & communication technologies	Annual plan	Acquisition process plan	Maintenance plan	Standards to be met plan	Training plan	R&D plan
Community Health Services						
Protection technologies	Annual plan	Acquisition process plan	Storage & distribution plan	Standards to be met plan	Training plan	R&D plan
Prevention technologies	Annual plan	Acquisition process plan	Storage & distribution plan	Standards to be met plan	Training plan	R&D plan
Promotion technologies	Annual plan	Hiring process plan	Distribution plan	Standards to be met plan	Training plan	R&D plan
Environmental Health	Master plan	Hiring process plan	Maintenance plan	Standards to be met plan	Training plan	R&D plan

the planners is specialized, however in many countries the planning is still in charge of administrators which can explain the weakness in the management of the technology and therefore the low quality of the technological environment that in turn leads to a low quality of the health service. Table 4.2 also shows the strength of considering all these aspects in planning, which obviously will depend on the size and level of the hospital, but on the whole shows the planning tasks we must do to ensure an appropriate technological environment. In this regard, it is desirable that in some countries, complete units of science and technology are being created within the health sector, generating possibilities for coordinated and integrated management of all aspects of technology. Likewise, planning may have a macro focus such as covering states or the country itself, or having a middle focus for regions or cities, as well as having a microfocus for planning within hospitals.

The products of the planning are reports written with objectivity, with justified affirmations that are the result of a specialized study in technologies for the health. Here are examples of planning products:

1. Institutional policies in health technology
2. Situational study of health technology with indicators
3. Institutional strategic plan in health technology with goals and indicators
4. Annual operational plan of the unit in charge of technology or clinical services, with goals and indicators
5. Master plan or multi-year plan for health technology, whether of the hospital as a whole or of each clinical service which should include all types of technology such as infrastructure, energy systems, communications, medical devices, clinical procedures, and organization
6. Annual procurement plan according to type of technology (see Chapter 1: Healthcare Technology Management (HTM) & Healthcare Technology Assessment (HTA)), whether new, replacement, consumables and/or spare parts as appropriate
7. Complementary annual procurement plan to health technology similar to the previous one and carried out very short term if necessary

In addition, it is necessary to plan the main activities of the unit in charge of the technology in the hospital and these will be detailed in the following chapters but here are some examples:

1. Plan to improve the organization and processes of health technology management (planning, acquisition, property or asset management, risk management, human resource development and research)
2. Plan to recover operability or reliability of medical devices, used especially when the hospital is in critical condition
3. Annual preventive maintenance plan and schedule, and criteria for corrective maintenance

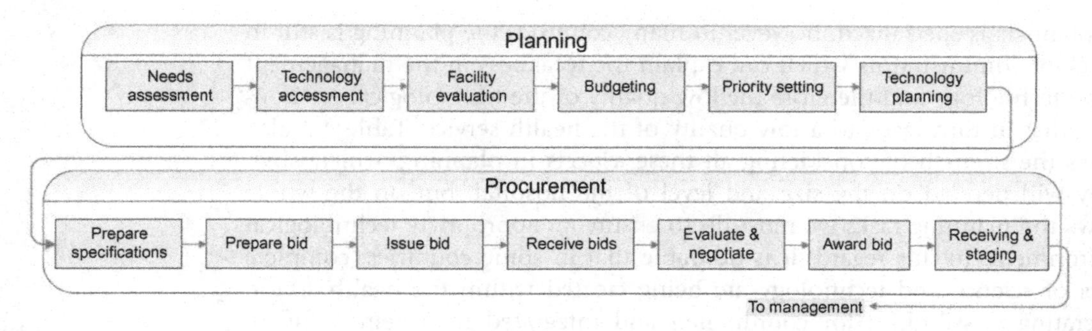

FIGURE 4.1
Planning and acquisition processes in a holistic HTM system. *Modified of ECRI.*

4. Annual risk reduction plan, which may include the hospital as a whole or each clinical service
5. Triennial or annual research plan and its financing
6. Triennial or annual health technology education plan to the hospital personnel
7. Training plan, interned and resided, in health technology, including supervised preprofessional practices for engineers and physicists

The main planning and acquisition processes are shown in Fig 4.1 which has been modified from ECRI, originally oriented to medical equipment. Nevertheless it can be a guide to be applied to other types of technology (Fig 1.1). The following subchapters will detail the planning and acquisition/procurement processes.

4.2 IDENTIFYING TECHNOLOGY FOR PEOPLE'S HEALTH

4.2.1 Technology Needs Analysis for Clinical Requirements

This is a planning process that tries not only to identify new technology for acquisition but also to improve what exists, that is, to adapt to the current requirements in terms of state of the art, for example in clinical procedures or resolving capacity, to make remodels or replacements. The needs by technology appear since the role and goals in health are defined for the hospital, and take into consideration the criteria of local epidemiology, demand for health services, technology incorporation procedure, project characteristic if any, technology with greater clinical effectiveness and/or lower risks, investment, and operating costs. All with the purpose of evaluating the health benefit for society and thus contribute to an investment or expense orderly and therefore to optimize the use of the resources. At this point the focus is more on technologies for individual and community health services (see Fig. 1.1) than on support technologies, and these latter will be addressed in the following stages.

A good example of a result or product of the needs analysis, in this case exclusively for medical equipment, is given by CENETEC (National Center for Health Technology Excellence in Mexico) [2], where it is proposed to issue a "Certificate of Need," similar to authorization procedure in building projects where the requirements of hospitals are "made official." Following this concept, we propose here that the certificate of need can also be applied by the unit responsible for the technology inside the hospital and not only for medical equipment, but for all types of health technology that, due to their relevance, require a greater analysis, So the Certificate of Need for Technology is requested by the clinical service or the medical branch in conjunction with the administrative branch, and is issued by the branch of technology:

- The requested technology should be included in hospital planning documents and correspond to catalogs or listings regulated by the health sector.
- The technology must be described or included in the master plan of the hospital or clinical service. Normally these plans are registered in the regional or national units as infrastructure master plan, which should change to technology master plan.
- Depending on the size of the technology, the application must be signed by the person responsible for the health sector, national or regional, or by the director of the hospital, and/or by the director of the clinical service. The request should be addressed to those responsible for the technology.
- It is necessary to have a template for better uniformity and facilitate the process of evaluating the need for technology.
- The application document must attach information validating the clinical need according to the role of the hospital in the health sector, as well as the calculation of supply and demand, map of the services network, epidemiological projections, etc.

The needs analysis for medical equipment is very well described in the link CENETEC [2] or publications such as [3,4]. However, the needs analysis for technology is broader and more complex, since it must focus on functional clinical service, that is, on operational, safe, with appropriate clinical effectiveness, supervised efficiency, and control of technology costs, such that clinical work and administrative of that service is carried out within the framework of effective quality. Any needs analysis begins by analyzing the situational state of the existing technology, in order to determine new technology requirements, it's according to clinical need or replacements or withdrawals of the service. This process should be designed to have a formal protocol, with formats, templates, and criteria coming from institutional policies, according to the vision and mission of the hospital. Replacement criteria should be clearly defined and justifications should have sufficient evidence, since it will be necessary to define priorities and cuts between several requirements according to the available budget items. It is important to valuing the clinical work and the technological possibilities that can be part of an institutional strategy and thus take advantage of financing opportunities, for example in case of having a high visibility in the market or positive marketing factor. Identifying technologies for clinical needs is one of the most interesting tasks for biomedical engineers and clinical engineers. This is a joint work with the medical assistance staff and should be in charge of those responsible for the technology in the hospital. Its results will impact in the short, medium, and long term, whether in the same clinical services, in the institution's economy, and especially in patients and users of the technology such as medical care personnel. Other engineering specialists such as electrical, electronic, and mechanical may not see this type of work. Likewise, this work can be systematized by elaborating questionnaires and interviews with the medical assistance staff, as well as officials or experts from inside and outside the hospital, since it is necessary to evaluate all types of impacts, among them the dimensioning of the technology required for the number of patients expected.

The review of the technology requirements for needs analysis can be made considering the criteria given in Fig. 4.2:

- Evaluating medical practice, in order to deduce the technological requirements, is a relevant task and does not always follow what is

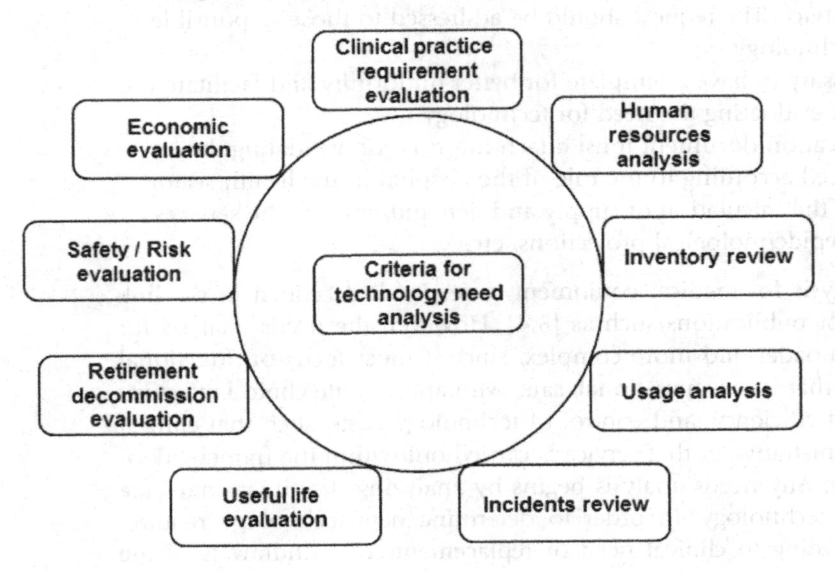

FIGURE 4.2
Criteria for technology needs analysis.

established in medical publications or regulatory entities for healthcare standards. Whenever a new technology allows for more safety, or is less invasive, or provides greater clinical effectiveness, medical standards advance but do not always keep pace with medical practice in the hospital, so the technology needs assessment can update the recognized standard if it had or should recognize the medical practice achieved.

- Consider the recruitment and retention of physicians due to up-to-date technology. A state-of-the-art technological environment is often appealing to specialist physicians who would not want to be held in an obsolete hospital.
- Review the catalogs, cadastres, and inventory of the various types of existing technologies. It is clear that the unit responsible for the technology must have implemented the means to have this information updated. Some technologies require a greater contribution of the medical branch, others of the administrative branch, and others are own of the technology in health. For the latter already has been suggested a coordinated work with professionals of engineering, architecture, physicists, biologists, among others. For example clinical procedures, drugs, and radiology require the input of specialists not engineers but under the responsibility of the branch of technology.
- Compare utilization statistics with the capabilities or time required by the procedure of existing technologies. This correlation must reach a point of equilibrium and be adapted to the current and projected requirements.
- Review the report of incidents related to each type of technology in use. For example, in the case of medical equipment there is an inventory and registration of work orders or maintenance databases, which can indicate the reliability and relative costs of the technology in use. In the case of energy systems, work orders or the application of new regulations or standards are reviewed. Clinical procedures can also be reviewed according to recovery times, clinical effectiveness, costs of supplies used, etc.
- Evaluate the useful life of the technology which involves considering not only the ideal estimates by manufacturer or the entity generating statistical data (e.g., 7 years or while having original spare parts or until the next version), but also multiple factors such as Frequency of use, reliability, design, supplier or manufacturer support, and computer platform upgrades. Advances in technology in terms of capabilities, multiple applications, and feasibility of commissioning can make existing technologies obsolete.
- Evaluating the retirement-decommissioning or replacement of technology requires clinical and technological expertise. There are many reasons for this decision, such as technological obsolescence, more

effective and/or safe technology, reduction of time and/or supplies, useful life, economic factors, among others. It will be necessary to define clear criteria for making decisions.

- Evaluating the safety of technology is a task that can be complicated because not all hospitals establish processes for the measurement of risks and thus generate actions for a safe environment. In some cases, security will justify the replacement or updating of a part or all of the technological products in use; in other cases, the detection of unsafe conditions, or the reporting of clinical or technological literature, or information alerts from the manufacturer or health institutions, or the hospital's own risk measurements, will allow for evidence-based decision making. It should also assess environmental and surroundings risks such as biological contamination, waste disposal, and the effect of ionizing and nonionizing radiation. In addition, risks related to natural and industrial disasters, which correspond to official entities such as civil defense and occupational health. They can be better addressed, as appropriate, by a hospital's branch of technology.
- Evaluating the economic factors of technology in health is one of the most relevant ones for the needs analysis since everything said already have a connotation in terms of costs. On the one hand there is the cost of incorporating a new technology, which implies not only the cost of purchasing but also the cost for its operation, such as supplies, licenses, personnel training, space, upgrades, installation, spare parts, and accessories. On the other hand there is the cost of operating the existing technology, which for the reasons described above may justify the withdrawal or replacement of a technology. Likewise, the economic factors should be evaluated considering the appropriate technology. In many cases there is a tendency to define a technological requirement with high performance or high sophistication, which is more expensive, for a requirement that corresponds to a lower level of health service, so search for balance and technological trends is one of the most challenging tasks. In the following subchapters the economic issue will be approached from another point of view as to the feasibility for decision made.

4.2.2 Healthcare Technology Assessment for Planning

As stated in Chapter 1, Healthcare Technology Management (HTM) & Healthcare Technology Assessment (HTA), according to the European Network for Health Technology Assessment—EUnetHTA [5], Health Technology Assessment—HTA is a multidisciplinary process that summarizes information about the medical, social, economic, and ethical issues related to health technology use in a systematic, transparent, and unbiased manner.

Its aim is to inform to make decisions considering safety, effectiveness, and health policies focused on patient and seek to achieve the best value to health services.

To assess technology needs some methods of HTA may be used by identifying requirements for the existing technologies (e.g., replacements or upgrades) or requirements for new technologies for the proper conduct of clinical procedures. HTA can provide information on technologies in the market that could be useful for the hospital. As discussed in Chapter 1, Healthcare Technology Management (HTM) & Healthcare Technology Assessment (HTA), HTA determines clinical effectiveness (in real conditions), clinical efficacy (under ideal conditions), risk–benefit ratio, cost–effectiveness ratio, as well as other relevant information statistics such as variability of clinical practice, meta-analysis studies for determination of best medical practices or ranges of doses or variables. It is about having information from the previous internal and/or external studies for evidence-based decision making, in this case for technology planning purposes.

Evidences for HTA come from publications on studies conducted by governments, HTA agencies, health research institutes, universities, etc., and also by studies done into the hospital or in the hospital network to which it belongs as part of its research role. In this point it is interesting to know the experiences and results of the technologies used in other places, be their countries or groups of countries, provinces, or other national hospital networks (macro-HTA), and also the experiences and results of the technologies used into the hospital (micro-HTA). The benefits of performing HTA are to meet the health needs of the attached population, reduce costs due to improvements in productivity and avoid the incorporation of unnecessary or obsolete technology, improve the quality and safety of health services, helps to have technologies that are consistent with the mission and level of service, have relevant information for further analysis to be carried out in the following processes of the HTM system. The purposes of HTA for planning are as follows:

- Prevent the adoption of untested technology that may not be clinically effective, cost effective neither safe.
- Maintain the appropriate order of priorities in the goals and health services delivery at the national and regional levels.
- Have an appropriate order of priorities for establishing clinical services within a geographical area based on an appropriate distribution of technological resources.
- Determine if the technology used in the hospital is appropriate and desirable for clinical practice and for the epidemiological needs of the patient population.

- Evaluate the technology provider and determine if the technology comes from the reliable source.
- Raise the final decision on the technology to be adopted based on the information collected and the conclusions of previous external and own studies. Comparisons between alternative technologies are very appropriate for this analysis.
- Plan the reevaluation of the technology after it has been installed, as well as the processes that were adopted or modified due to the incorporation of this technology. It is necessary to know if the technology adopted has met expectations in improving the quality of health services, effectiveness, safety, expected costs, and patient stay times. It is also necessary to know whether we should make modifications to the following technology profiles, or whether our own evaluation process should be improved for future reliability. Having a branch in charge of technology favors this work.

4.3 EVALUATING SUPPORT TECHNOLOGY AND FINANCIAL FEASIBILITY

Following the planning process, after identifying the technology needs required for health services and identifying the recognized and appropriate technologies alternatives in the market, it is time to carry out an analysis of the feasibility of the proposals in terms of their impact on facilities, organization, and economic impact on the hospital. The information generated in this part of the process will allow to define priorities and in addition the budget before going to the procurement process. Under the concept of our model of technology branch, which involves a collaborative work with other specialists but with responsibility in technology, according to the scope of the project, it is clear that there will be interaction with architects of hospitals, hospital engineers, electrical engineers, mechanical engineers, computer engineers, electronic engineers, medical physicists, biomedical technicians, etc., as well as naturally the interaction with medical specialists in its various branches, nurses, medical technologists, biologists, chemists-pharmacists, etc., and specialists in administration, logistics, officials, etc.

Clinical services with a functional technological environment require that all types of technology work simultaneously and in a coordinated manner. In the technology needs analysis for health services, technologies for individual and community health services were more closely addressed. In order to complete the feasibility of the alternatives found, we will now discuss the required support technologies (see Fig 1.1) and financial evaluation. In addition, if it is a new health facility, which includes all types of technology, Section 4.4 has been considered.

4.3.1 Facility Evaluation

Table 4.2 shows examples of planning products for each type of technology. In the case of support technologies and following our holistic approach, not only must infrastructure and energy systems be considered for the operation of technologies chosen for individual and community health services, but also what is needed for information and communication systems, the impacts on the organization and processes of the clinical service due to the incorporation of a technology should be even more foreseen. It will be necessary to determine if the clinical service has the human resources and physical resources sufficient to assimilate and support a new technology.

In the case of medical equipment, there is abundant information that explains their infrastructure requirements, electrical energy, environmental conditions, water, maintenance human resources, etc. The complexity in this case is in the number of equipment to be installed and the diversity of equipment that must operate in a coordinated way in a single environment. For this, a design or redesign of the clinical service is required to properly distribute the workstations according to the current norms and regulations. Also, it should be considered that medical devices are means or tools for performing clinical procedures, but how are clinical procedures managed? these are a type of technology that requires a plan to incorporate, update, train, and evaluate their effectiveness, and even a research plan to improve or generate new clinical procedures. Also, drugs and other consumables require support technologies. Some require storage under appropriate environmental conditions and also require various information systems, such as inventory management, distribution, patient supply in clinical services, etc. New technologies for drug management need to be evaluated as robots for medicines delivery, as well as technology for the management of radio-drugs, among other alternatives. A branch of technology can better meet these requirements.

One of the most neglected aspects in hospitals is the organization and management system associated with the incorporation of a health technology. And is that even with the existing technology, the impact of these is not always clear given the lack of a complete management system with processes, indicators, and responsibilities. The management system for clinical services is an issue that should be reviewed by all actors, medical care personnel, administrators, and engineers. A training plan and applied research plan can help design better processes, better organization and evaluate their efficiency and productivity. The incorporation of new technology brings new tasks, additional procedures, new distribution of supplies and medicines, as well as new or different circulations, and new distribution of responsibilities. It is true that the head of department or clinical department is responsible overall, but we must assign responsibilities for the technological as is done with the administrative, so we return to our approach that any successful health intervention requires considering the clinical, administrative, and technological focus.

4.3.2 Financial Evaluation

Life cycle cost (LCC) analysis can be applied to determine the cost of ownership of a technology, e.g., the cost of ownership as long as the technology remains in the hospital. This life cycle has been called in Chapter 1, Healthcare Technology Management (HTM) & Healthcare Technology Assessment (HTA), the healthcare technology application cycle in a hospital (Fig 1.6). All costs associated with the technology acquisition and operation are considered during the time of use in the hospital. Some estimate this time in an average of 7 years for medical equipment, although some other technologies like software can be of only 1 year and the hospital infrastructure can be projected to 40 years or more, thus the knowledge of the own and external antecedents will provide the best information for each type of technology. The costs to be considered are purchase price, insurance, transportation, installation, authorizations or licenses, consumables and disposables, maintenance, spare parts, upgrades, training, new personnel, as well as financial aspects such as inflation, taxes, and discounts. In many cases this information is not available so it will be necessary to make comparisons and estimates for planning purposes. For example, benchmark pricing would require having up-to-date databases that can be obtained by taking information from other hospitals or ECRI. However some adjustments are necessary to consider local premises and estimate the total cost. A technology branch could raise policy in this regard and help significantly to have this type of information relevant to planning.

On the other hand, it is necessary to indicate the cost of not doing what is appropriate. Not very well managed technologies consume wasted resources, especially economic resources that can well be used for other requirements. For example the possible energy savings to be achieved, such as electricity, oxygen, steam, should be considered in planning. Also, the lack of use or poor operation of some technologies must be reviewed before generating waste of economic resources. This is one more reason to have a responsible branch of technology in a hospital. Some experiences in hospitals in developing countries have shown that the savings far outweigh the investment made, for example, on hire additional specialized engineers.

4.4 PLANNING OF NEW FACILITIES OR RENOVATION

One of the major planning challenges is when it comes to setting up a new facility or renovation such as a new hospital, a new clinical service, or the renovation of an existing clinical service. There is sufficient information regarding the design and construction of new facilities, but we would like to point out our holistic view here to ensure the functionality of the technological environment in the clinical service. The concept of designing the architectural component and then proposing the equipment is moving away from reality. If the end product is the functional technological environment, all types of health technology must be considered in the early stages of design, so the key to a successful project is to maximize planning in the early stages to minimize changes throughout the process. For this purpose, the design team consists of the project manager, hospital architect, and engineers of various specialties, and on the other side, the user team of the clinical service or hospital, it is made up by medical directors, administrators, and the staff of health professionals along with security specialists, communicators in health, interior designer, etc. In addition, subcommittees are formed to address specific issues such as clinical engineering, technology management, communication networks, infection control, and security. It is important to note that biomedical engineers and clinical engineers may belong to any of the aforementioned teams or subcommittees due to their training not only in technology but in physiology and clinical procedures.

The construction processes of hospital facilities usually take the following steps: Planning, schematic design, design and development, construction documents, construction, and commissioning [6]. The planning stage is one of the most critical, which includes the purposes of the clinical service or hospital, the "wish list" of how is visualized the operation of the clinical service. Then master plan will be developed and predesign alternatives are proposed. Usually a recognized firm in architecture and engineering is hired for this work, but we want to draw attention to the fact that the hospital should naturally be involved not only in the participation of its medical and administrative medical staff, but also must involve the personnel responsible for the technology.

In this case, planning begins considering the approaches given in institutional documents, strategic plan, vision and mission, roles, goals to be achieved, business plan, etc. The expected improvements in services and patient satisfaction, medical care staff, and visitors are also considered. The health needs of the community are also identified and incorporated. It is also very important to consider the technology that will be procured, and this is the point that should be highlighted, because as we know, there are different types of technology that should form a coordinated functional environment, but not always the infrastructure can accommodate each type of technology, more and more technologies condition the infrastructure, and this is the analysis in which biomedical engineers and clinical engineers must be involved in order to meet the clinical requirements with effectiveness, efficiency, safety, controlled costs, and quality of service. Environmental requirements should also be considered in order to make them compatible with people and with the technologies incorporated. The topics of biological contamination, radiations, gases, etc., become relevant, as well as electromagnetic compatibility, interference between equipment, etc. Circulations of medical care personnel, medicines, medical materials, waste, and others are part of these initial considerations. A branch in charge of the technology is very useful in this whole process because it naturally becomes an articulating counterpart of the hospital to the contracted companies.

With the information collected and the initial analysis carried out, the elaboration of the master plan and the predesign is continued. To do this, the needs analysis information of the new clinical service or new hospital is used, the working teams and committees are formed and more data is collected. The information is consolidated in an establishment program (facility program) in order to request the certificate of necessity and authorization, if applicable. Then the activities are scheduled and the budget for all stages is created. The consolidation of all these documents constitutes the final document of the master plan and predesign, which will allow going to the following design stages of building and construction. The master plan should also consider a series of regulations and standards such as safety, emergency management, patient flow, life safety, infection prevention and control, and proactive risk assessment before construction. Our position in the holistic view is that this master plan should contain the planning of all types of technology to achieve the appropriate functionality of the clinical service or hospital, and this is the result of the participation of a group of specialists among whom are the responsible for hospital technology such as clinical engineers, biomedical engineers, medical physicists, among others.

4.5 PRIORITY SETTING AND BUDGETING

Knowing that the resources are limited and the demands are usually unlimited, several requirements will not be financed neither attended. Thus, the budget becomes a political document that the institution will adopt in an institutional framework of policies, priorities, resources and uses, goals, opportunities, organizational structure, criteria, etc. Regardless of legislative, economic, or environmental constraints, the hospital will use its resources to achieve its goals and objectives. We observe that budgets are usually similar over the years, they are applied equally in lean or in apogee moments, the budgetary goals are adopted publicly, nevertheless one must consider the more formal methods, that is to say, coherent and defined criteria, Transparent for their better understanding and confidence, and with more supporting information. A committee should be formed to establish priorities with the participation of the units involved and authorities of the institution.

The key components for setting priorities are the following:

- *Education*: The trained staff will have a better understanding of the contents and approaches, and thus assess the process of setting priorities and for what purpose. If the problems are correctly explained, together with the challenges, parameters, resources, and demands, it will be possible to identify the desired product or result. Context information and outlook will help to prioritize regarding other requirements.
- *Communication*: During the process of establishing priorities, it will be necessary to maintain an effective communication with the interested units in order to receive information that can be used in the process, as well as to provide information that compiles priorities, giving equal opportunity to all.
- *Framework*: To have a structure that helps to prioritize according to the objectives, considering the clinical service category, beneficiary category, geographic location, stakeholders, goals with priorities, as well as to provide the rules of the process with respect to those who participate, desired results, expectations, and timeline for the process and priorities.

Priority setting refers to choosing between problems and related solutions, so it will be necessary to identify and prioritize problems in addition to identifying and prioritizing solutions. When a problem is identified, it will be possible to visualize a result or goal, for which we can assign objectives and therefore define the activities or steps to follow to achieve that result. In the final discussion, we try to prioritize problems and solutions considering influential factors:

- Financial resources—How much does it cost?
- Human resources—How many people are needed to implement it?
- Other resources—What other resources do need to be dedicated to this?
- Legislative impact—Does it require changes in local, state or regional law, guidelines, or policies?
- Impact on other operations—Do other services have to be adjusted to make this happen?
- Educational needs: What kind of publicity or public education will be needed to implement it?
- Affected parties: Have everyone been identified and, if intended, have they been contacted?
- Affected parties: How many people, especially relative to the total possible, will be affected?

- Affected parties and alternatives: How critical is this to the affected parties' health, safety, and wellbeing?
- Timeframe: How quickly will this affect stakeholders?

Factors such as these can be weighted and can be assigned a quantitative rating. The points assigned to each factor will allow the end to evaluate the best alternatives. This is a simple, accurate, and fast way for all interested parties to understand the decision made. With the prioritization made, it is possible to allocate the amounts previously established to each item to form the final budget.

4.6 ACQUISITION PROCESS OF HEALTHCARE TECHNOLOGY

The processes of acquisition of health technologies are fairly well documented individually, i.e., there is enough information to guide the process of acquiring medical equipment, as well as for medical materials or to hire health facility building, etc. However, if we consider that there is a unit in charge of the technology in a hospital, it will be necessary to gather all these capacities in order to manage them in a more coordinated and integrated way so that a functional technological environment for the clinical service or hospital is reached. It is known that the acquisition process must comply with the local regulations and conditions, the current procedures for the acquisition method, the legal and administrative framework, etc. but how do we have access to up-to-date clinical procedures? How do we buy prevention technologies? How do we close the contract for a remodeling that will affect other units? These knowledge and practices exist but in a dispersed way and we think that they must be gathered under a single organization in charge of the technology, that is to say the technology branch of the hospital.

Examples of the procurement process for each type of technology have been compiled in Table 4.3, the first stage is common to all, it is necessary to prepare the specifications in order to achieve the requirement in the market, this is a key process that we will develop in the following lines; then all kinds of technology require that a call for proposals be prepared for bidders to propose their products. The call can be as simple as requesting a quote or as complex as launching an international tender, in all cases the calls are described in official documents where rules, requirements, deadlines, reference amounts, conditions, etc., are set forth. Both public and private entities have formally regulated criteria and others specific to each entity. The offers are received and a committee specially trained for this stage evaluates both clinical and technological aspects as well as economic ones. The evaluation can be simple with few items or very complex with hundreds of items to buy with a great diversity of suppliers. After the evaluation of the offers, the

Table 4.3 Examples of Technology Procurement process in HTM System According to Type of Health Technology

Technologies for Healthcare	Clinical Service or All Hospital						
Individual Health Services							
a. Clinical technologies							
Clinical procedures	Prepare specification	Search process	Ask for quotation	Receive quotation	Evaluate & negotiate	Suscription or contract	Receiving & delivery to users
Medical devices	Prepare specification	Prepare bid	Issue bid	Receive bids	Evaluate & negotiate	Awards bids	Receiving & staging
Pharmaceuticals & others consumable products	Prepare specification	Prepare bid	Issue bid	Receive bids	Evaluate & negotiate	Awards bids	Receiving, storage & distribution
b. Support technologies:							
Organization & management system	Prepare specification	Search process	Ask for quotation	Receive quotation	Evaluate & negotiate	Sign contract	Implementing & staging
Infrastructure & hospital equipment	Prepare specification	Prepare bid	Issue bid	Receive bids	Evaluate & negotiate	Awards bids	Receiving & staging
Information & communication technologies	Prepare specification	Prepare bid	Issue bid	Receive bids	Evaluate & negotiate	Awards bids	Receiving & staging
Community Health Services							
Protection technologies	Prepare specification	Prepare bid	Issue bid	Receive bids	Evaluate & negotiate	Awards bids	Receiving, storage & distribution
Prevention technologies	Prepare specification	Prepare bid	Issue bid	Receive bids	Evaluate & negotiate	Awards bids	Receiving, storage & distribution
Promotion technologies	Prepare specification	Prepare bid	Issue bid	Receive bids	Evaluate & negotiate	Sign contract	Implementing & staging
Environmental Health	Prepare specification	Prepare bid	Issue bid	Receive bids	Evaluate & negotiate	Awards bids	Receiving & staging

purchase is awarded to those who have the best scores. The process continues with the signing of the contracts and ends with the reception, installation, and commissioning of the products purchased. There are some advantages that an integrated management system HTM or technology branch can do, for example to carry out a coordinated programming of each type of technology and of items to buy, so that a deadline is reached with everything installed and functional, at the moment it is very difficult to do something like that since there is not a single criterion responsible for the technology, each type of technology is in charge of a different unit. Even when the hospital hires a company for a turn-key purchase, an integrated HTM management system can be very useful for jointly monitoring the progress of work by considering all types of technology.

Since many of the processes of acquisition are already regulated and described in their places of origin, we will not repeat here, what we do want to emphasize is with respect to the technical specifications. WHO has very good publications of technical specifications for medical equipment [7,8] and other similar ones. The technical specifications are official formats developed for purposes of comparison between several technological alternatives that allow an objective evaluation to find the best proposal. These are or should be the result of needs analysis, health technology assessment, and referral product costs. They require continuous updation given the progress and permanent changes of the technology, but not to achieve the most sophisticated, but the most effective, safe, or efficient, for that constitutes a continuous work of handling specialized information. Clinical practice guidelines do not adequately describe the specifications of the required technology, but it is also true that some new technologies make it possible to improve clinical practice guidelines.

The following are some recommendations for the elaboration of technical specifications of the technologies required by health services or hospital:

- Name, category, and coding: Use standard or current nomenclature.
- Purpose of use: Specify clinical purpose, type of health service, and level of use.
- Technical characteristics: Describe the functional requirements, relevant components, range and precision of the variables, display functions, alarms, and configurable parameters.
- Physical/chemical characteristics: Specify type of materials, dimensions, weights, configuration, mobility, etc.
- Utility requirements: Dimensioned requirements of electricity, gas, water, drainage, etc.
- Accessories, consumables, spare parts, and other components: Accessories required for the place of operation, method of sterilization, number or capacity of consumables, renewable or disposable consumables.

- Packaging: State of sterilization, conditions of transport and storage, labeling.
- Environmental requirements: Ambient operating conditions, temperature, humidity, altitude.
- Warranty and maintenance: Warranty conditions, warranty period, conditions for maintenance, metrological verification and updates, availability of spare parts after warranty.
- Documentation: Operation and service manuals, with list of spare parts and consumables, with description of metrological verification procedures.
- Decommissioning: Indicate estimated time of life or validity.
- Safety and standards: Risk classification, approval of external and local regulatory agencies, compliance with local and international standards.

In the following chapters, the processes of asset management, risk management, human resources development, and healthcare technology research will be developed, and then we will discuss proposed organizational models for a technology branch of hospitals.

References

[1] Medical equipment planning. Health devices, vol. 26, no 1; 1997. p. 11, 12.

[2] Guide 2: how to plan and budget for your healthcare technology. Ziken International. 'How to manage' series for healthcare technology. Series Editor: Caroline Temple-Bird. UK. 2005.

[3] WHO. Needs assessment for medical devices. WHO Medical device technical series; 2011.

[4] Guide 2: how to plan and budget for your healthcare technology.

[5] European Network for Health Technology Assessment—EUnetHTA. <http://www.eunethta.eu/>.

[6] Joint Commission Resources (JCR). Planning, design, and construction of health care facilities. 2nd ed. Oakbrook Terrace: Joint Commission on Accreditation of Healthcare Organizations; 2009.

[7] WHO. Technical specification for medical devices. 2014. Ziken International. 'How to Manage' series for healthcare technology. Series Editor: Caroline Temple-Bird, UK; 2005.

[8] Guide 3: how to procure and commission your healthcare technology. Ziken International. 'How to manage' series for healthcare technology. Series Editor: Caroline Temple-Bird. UK. 2005.

Further Reading

Guide 1: how to organize a system of healthcare technology management.

Guide 4: how to operate your healthcare technology effectively and safely.

Guide 5: how to organize the maintenance of your healthcare technology.
Guide 6: how to manage the finances of your healthcare technology management teams.
WHO. Procurement process resource guide. WHO Medical device technical series, 2011.

CHAPTER 5

Asset & Risk Management Related to Healthcare Technology

The way of seeing maintenance and equipment management must evolve to observe improvements in the hospital technological environment.

Following our premise of what is the functional technological environment that qualifies the result of the work of an engineer in hospitals, be with clinical engineer or biomedical engineer profile, and that health technology is not only the medical equipment, but a series of components more fully described in Chapter 1, Healthcare Technology Management (HTM) & Healthcare Technology Assessment (HTA), which includes:

- Technologies for individual health:
  - Clinical technologies:
    - Clinical procedures
    - Medical devices
    - Drugs and medical materials
  - Assistive technologies:
    - Organization
    - Information & communication technologies
    - Hospital equipment & energy systems
    - Infrastructure
- Technologies for community health:
  - Protection technologies
  - Prevention technologies
  - Promotion technologies
- Technologies for environmental

And among these, there are hard technologies but also soft technologies such as the organization, its processes and methods, clinical practice guidelines, and information systems, all this invites us to think about technology in health with a holistic approach, more oriented toward the purposes of the hospital and therefore focused on the patient. Naturally, such an approach will motivate more engineers to work in hospitals, as long as hospitals tailor

Healthcare Technology Management Systems. DOI: http://dx.doi.org/10.1016/B978-0-12-811431-5.00005-9
© 2017 Elsevier Inc. All rights reserved.

their organization to benefit themselves and open up spaces for the broad and deep work of engineering. To a great extent, this is the motivation of this book, to create a new model or, in any case, to improve the current models of hospital organization in order to have the real capacity to manage its own technology.

Chapter 4, Health Technology Planning and Acquisition, discussed the planning and acquisition of health technology, following the cycle of application of technology in hospitals (hospital life cycle). It is now appropriate to address the application or use or operation of this technology for the purposes for which it was acquired. The products or technological goods together and in a coordinated manner conform the technological environment in hospitals and systematic actions are required to guarantee their functionality in a framework of continuous operation, safe, efficient, cost-controlled, and more clinically effective. By fulfilling these characteristics we reach the quality levels expected from the technology point of view, which predisposes better levels of quality of healthcare services to the patients. Is it possible to provide quality health services only with good health professionals? Our premise says that all health interventions involve three components: medical care, administrative support, and technology support. Any weakness in any of these leads to deficiencies of the health service provided. Therefore, the quality of health services is supported by appropriate health professionals (medical branch), proactive administrative branch, and equally in the quality of technology used (branch of technology).

In this chapter we intend to demonstrate that hospital engineers should not only focus on the maintenance of medical equipment or equipment management, but their work should be expanded to the point of considering management for the functionality of the technological environment of clinical services as a whole. This approach will greatly aid the health sector and the work of health professionals, as well as being more compatible with the training of clinical engineers and biomedical engineers, who stand out for their knowledge of physiology and technology at the same time. Also, this chapter addresses the issue of document management and information systems, often forgotten or relegated, but which allows the real capacity to manage the technology. We will also develop the operation of hard technology (machine centered) and soft technology (human centered) [1], as well as address inspections under our holistic approach to health technology. The inspections are not only to ensure the operability but also to reduce the risks in the use of the technology. Just another important aspect to be presented in this chapter is risk management, in many hospitals without an appropriate person who is responsible or with dispersed and incomplete responsibility in many persons. This point is one of the most urgent ones to apply and we deal in part with inspections, but also considering the techno surveillance, registration, and management of adverse events related to cases involving technology, rather than to sanction is for Improve the current conditions and that these events do not happen again. Also included is occupational health, which like any health intervention has technological aspects that must be managed. Metrological verification is also developed, which as a procedure can be considered within the asset management category, but due to its impact on the health of people through the information provided by medical devices for diagnosis and treatment, we have considered it to be part of risk management. In addition, within the framework of managing the technology from a single unit responsible for the hospital (technological branch), it is opportune to include topics such as the control of hospital acquired infections, hospital waste management, and disaster mitigation, which are faced with knowledge scientific and expertise based in technology.

5.1 MANAGING TECHNOLOGY ASSETS FOR CLINICAL SERVICES

To begin, we must review the evolution of hospitals, considered today as one of the most complex entities, as well as being a resource among the most important ones in a community. According to Coe [2] "A hospital is, above all, a place where members of the community can obtain services to restore their health. More recently it has also become a place for the rehabilitation of people physically incapacitated or where the elderly can recover. The modern hospital is also a teaching place, a learning center for future doctors, surgeons, and other professionals. At the same time, the hospital is often a research center where scientific knowledge of diseases is broadened. In a sociological sense the modern hospital is a large and complex organization, with a hierarchy of status and roles, rights and obligations, attitudes, values and ends." But also in the present times we have definitions like that of Frisch [3] "an intelligent hospital is one that works better and smarter. It's better because it's resourceful, creative, and perceptive about what patients and doctors need, and it's smarter because it's astute and inventive when it comes to weaving together diverse technologies to enhance patient care. Driven not only by new regulatory requirements but by financial constraints and reductions in staff that requires us to do more with less." The medical-patient encounter is not now essential, Information and communication technologies (ICT's) allow a remote encounter, and telemedicine is another technological component that will continue to be incorporated into health services. The history of technology in hospitals is intermingled with the history of medicine. The technology into hospitals is formally recognized from the early years of the 20th century with the introduction of the X-ray machine, although medical technology has existed for many centuries. So, in

1900, who was responsible for the technology? Usually the doctors and nurses themselves. There is not much information about the history of maintenance in hospitals and is that it appears as a repair office in the first half of the 20th century, something accessory but necessary, no one would have to pay much attention to it compared to the work of doctors and nurses. Where is its location? The basement so that it does not interrupt the "own" labors of the hospital. After the Second World War, with the rebuilding of the industry, the departments of maintenance arise, and of course the hospitals follow this premise. Almost 70 years later, this is still the model to attend to the operation of the equipment, which was the requirement at the time, but it is not the model to deal with the healthcare technology management (HTM). Maintenance departments are necessary for the operability of the equipment, but they are not sufficient for the technological development of the hospital, we require a new model with capacity for design, planning, asset management, risk management, human resources development (engineers and others), and research, and naturally with responsibility in its subject and with the leadership role to propose policies, strategies, to regulate and control the technology in the hospital.

The technological asset management and risk management due to technology are two other pillars for technology management in hospitals. Their processes happen in the day to day, that is to say they are more operative and therefore a good management must tend to the systematization of the procedures. The goal is to have a functional technological environment, this means not only that the technological components are operational individually, but also operative together, with reliability, efficiency, safety, and cost control, so as to ensure the expected clinical effectiveness of the clinical procedures that are provided to patients. The old model of maintenance or repair only raises the operability of medical equipment and many times individually, which is insufficient for the current circumstances. One justification for systematization is the number of technological components or items to be served. Depending on the size of the hospital, we can have hundreds and thousands of items, which virtually makes manual or paper handling impossible. In addition, given the high repeatability of the procedures, and to ensure the suitability of the work and its speed, it is necessary to have standard procedures (standard operative procedure, SOP). The design of SOP's is an interesting work for clinical engineers or biomedical engineers, and better if we consider the full range of technologies contained in a hospital, because diversity poses challenges appreciated by engineering professionals. SOPs can also be shared with other similar hospitals respecting property rights.

Fig. 5.1 shows an example of process map, including strategic, operative, and support process, that they should currently be considered for the technological assets management under a holistic perspective of technology. Naturally,

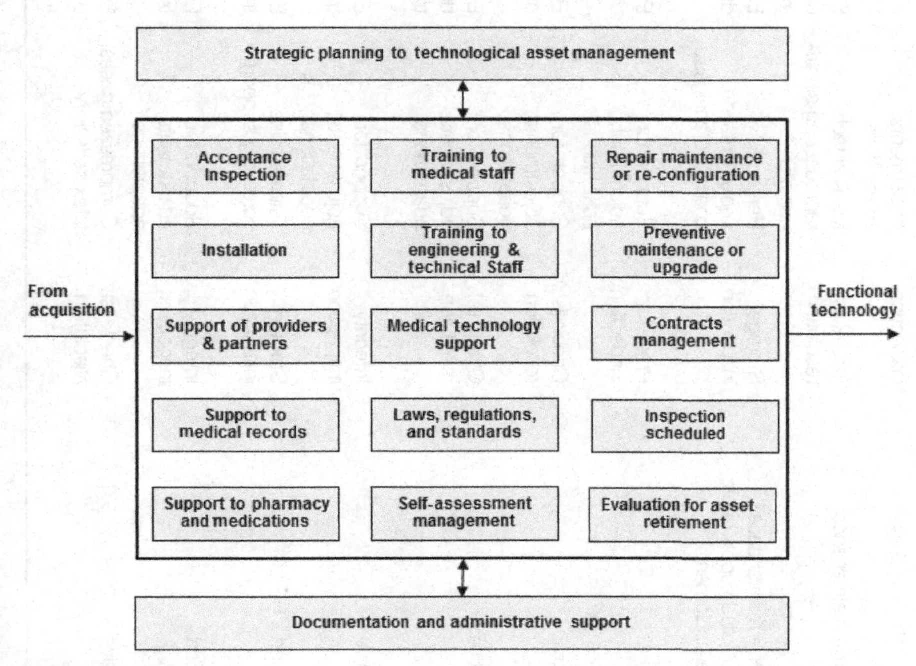

FIGURE 5.1
Example of process map to technological asset management in a hospital.

a suitable interpretation will be required for each type of technology. Also, the application of these processes to a specific type of clinical service will require appropriate adaptations and the inclusion of specific details according to the expertise achieved in the hospital.

The need for a reliable IT platform to properly handle processes is evident. Hospitals usually have a different computer support unit than engineering departments, however, under a holistic approach to technology management, both units can be incorporated into the hospital's technology branch and be more effective in achieving the goal of functional technological environment. Naturally, any organizational change is not easy to raise, especially in strong and well-established organizations, but we also have hospitals with a very weak organization for the management of technology, where this approach will be more viable and meaningful.

In Table 5.1, we have grouped the main processes for technological assets management in order to show examples of the typical actions to be performed in each one. The different types of technology have been considered for this purpose. However, in order to be more precise, the specific clinical service must also be considered. Therefore, Table 5.1 can be used as a

Table 5.1 Examples of Asset Management Actions According to Type of Health Technology and Clinical Service

Technologies for Healthcare	Clinical Service or All Hospital					
	Installation	Training	Maintenance	Inspection & Compliance	Support to Hospital	Retirement
Individual Health Services						
a. Clinical technologies						
Clinical procedures	Implementation	User training	Updates	Clinical inspection	Support for adjustments, adaptations	Effectiveness evaluation
Medical devices	Installation, Acceptance, Inspection	Users & Staff Support	Repairs, preventive maintenance	Check list, resolution	Medical tech., reprocessing single-use MD	Evaluation, disposition, transfer
Pharmaceuticals & others consumable products	Acceptance	Users & staff support	Repairs, preventive maintenance, temperature-controlled supply chain support, storage equipment	Storage inspection	Barcode implementation, controlled analgesia	Evaluation, disposition
b. Support technologies						
Organization & management system	Supervision of process incorporation	User training	Update	Process inspection	Support for adjustments, adaptations	Evaluation, disposition
Infrastructure & hospital equipment	Installation, acceptance, inspection	Users & staff support	Repairs, preventive	Check list, resolution	Support for adjustments, adaptations	Evaluation, disposition
Information & communication technologies	Installation, acceptance, inspection	Users & staff support	Repairs, preventive	Check list, resolution	Support for adjustments, adaptations	Evaluation, disposition, transfer
Community Health Services						
Protection technologies	Acceptance	Users & staff support	Repairs, preventive	Integrity inspection	Support for adjustments, adaptations	Evaluation, transfer
Prevention technologies	Acceptance	Users & staff Support	Cold chain support	Storage inspection	Telemedicine, screening support	Evaluation, transfer
Promotion technologies	Acceptance	Users & staff support	Cold chain support	Integrity inspection	Support for adjustments, adaptations	Evaluation, transfer
Environmental Health	Installation, acceptance, inspection	Users & staff support	Repairs, preventive	Check list, resolution	Compressed gases, latex sensitivity	Evaluation, disposition, transfer

template to be filled with the actions for each clinical service, according to characteristics and dimensions of each hospital. Given the large volume of tasks, we reiterate the need to systematize processes, rely on IT platforms, and prioritize the technologies to be addressed. In addition, not all technologies have the same lifetime, some are just days and others are long years. Also, some are highly complex and others of medium or low complexity, so it is a task of classification, starting with a standard nomenclature, the assignment of an identification code that determines type of technology, technological complexity, risk to people, etc. Additionally, tasks for clinical engineers or biomedical engineers are focused on designing, directing, or supervising. The diversity of tasks will allow assigning them to different profiles of training, technical, bachelors, master, or doctorate, while each engineer and technician may specialize according to the type of technology or more conveniently according to the technology for each type of clinical service.

Next, we will detail the main processes for technological assets management, as part of a HTM integrated system for hospitals. The training and educational aspects can be seen in Chapter 3, Human Resources and Healthcare Technology Workforce, and Chapter 6, Quality & Effectiveness Improvement in the Hospital: Achieving Sustained Outcomes.

5.2 DOCUMENT MANAGEMENT AND INFORMATION SYSTEMS FOR HEALTHCARE TECHNOLOGY

One of the great challenges for a HTM system is the management of your information, either virtual or in physical documents. This is also true for the technological asset management and risk management in particular. There are many flaws and shortcomings and in this respect the initiatives are often partial or incomplete. Thus, the compatibility of information between units is not always possible, they are not integrated, the information of doctors and patients (medical branch) does not follow the same pace as administration and logistics (administrative branch) and of course information technology (technology branch), usually referred only to medical equipment, is not always available to the point that does not allow the analysis of results, cross-checking of trends, or obtaining individualized costs, etc. Where are planning documents kept? and the work orders? Where are medical Equipment Manuals? the rules and regulations? the catalogs of parts or medical devices? books and magazines? Under the holistic approach to HTM, a strong and consistent information system is needed, in addition to document management capacity. HIS (health information systems) [4] or similarly HIT (health information technology) [5] are usually oriented to the health business process and focus on providing better health services, we refer to information system so that it can manage healthcare technology, and therefore it supports the information derived from the processes of HTM. There are software packages for the maintenance of medical equipment, some incorporate aspects of planning, pre-defined templates, cost management, etc., but do not extend to the management of other types of technology. This approach opens up an interesting space for software development in a more holistic version.

The management of the documentation and information system for HTM will take the concepts of library management and use appropriate software for information management. Of course, the size and complexity of the hospital will determine the dimensions of this capacity. A library with information system is a business center [6]. Some hospitals already have a library normally oriented to medicine and can be strengthened with the requirements for the technology management. In other cases it may be decided to have this capacity specifically for technology management. This library is composed of the following sections, based on the services rendered:

I. Acquisition Section: The books, manuals, standards, etc., demanded by technology branch should be considered in the operational budget. It is also useful to have agreements for the exchange of information with other libraries of hospitals or universities, and to have access to specialized databases.
II. Technical Section: To classify, catalog, online public access catalog (OPAC), barcode, etc. It's known that healthcare technology also has important advances, but hospitals do not have the means to receive this information because their organizational structure and functions are not given. One of the important aspects is to have the nomenclature, for example GMDM medical devices [7], the classification of diseases ICD-10, and the International Classification for Patient Safety (ICPS).
III. Circulation Section: This section provides documents to the members and users of library for home or office reading as well as reading in the library. The documents are arranged on the racks in stacks as per the Dewey Decimal Classification Scheme.
IV. Periodical Section: To search, maintain, and arrange periodicals services to library members. Periodical section provides reference service with respect to the requirement of the reader. Reference section is attached to periodical section for convenience of library users.
V. Databases: Databases like OPAC, World Health Organization (WHO), ECRI, and online network service are available for the library members. It includes the databases with information of the technology branch that may have restricted access for some users. It

will be convenient to have computer terminals for exhaustive searches of information.

VI. Reprographic Section: This section has two automatic plain paper copiers and a duplicating machine through which the photocopy services are provided to the readers. Depending on the case, this service can be contracted to a third party.

VII. Binding Section: This section is working for binding work of the damaged books. Back Volumes and other documents of this Library. The required binding service can be contracted.

5.2.1 Nomenclature, Inventory, and Work Orders

The importance of establishing and using a nomenclature in medical devices is given by [7,8], this is to facilitate management and regulation by means of standardized terms such as to allow effective communication despite language barriers and others as may be result of the multi-professional interaction. There is a high diversity of nomenclature systems developed by professional groups for their own purposes, such as maintenance, procurement, accounting, stock keeping, regulatory affairs (product registration for marketing authorization), medical device adverse event reporting, and customs operations. The WHO promotes a nomenclature system for globally unified medical devices: Global Medical Device Nomenclature (GMDN) [8], developed after a joint work with ECRI and its Universal Medical Device Nomenclature System (UMDNS). For the purposes of holistic HTM, it will be necessary to have an even more extensive nomenclature system than only medical equipment or hospital equipment, but also to consider nomenclature for different types of technology such as clinical procedures (with ICD-10), terms that describe the policies proposed, processes and procedures established (SOPs), with medicines we have consolidated nomenclatures, but still several developments are required for community health technologies.

The inventory is another essential element for the HTM. WHO defines it as "A medical equipment inventory provides a technical assessment of the technology on hand, giving details of the type and quantity of equipment and the current operating status" [9]. This definition should shift from inventory of medical equipment to inventory of technology for application in an integrated system of HTM. It is clear that you want to know the current status of the technology item, but also their associated events such as maintenance work orders, inspections, and alerts. Therefore, the inventory has the characteristic of being permanently updated. There is information that will not change in time (brand, model, specifications, etc.), but there is other information that is dynamic or cumulative over time and can generate specific database such as work orders that are cumulative, state of operation is

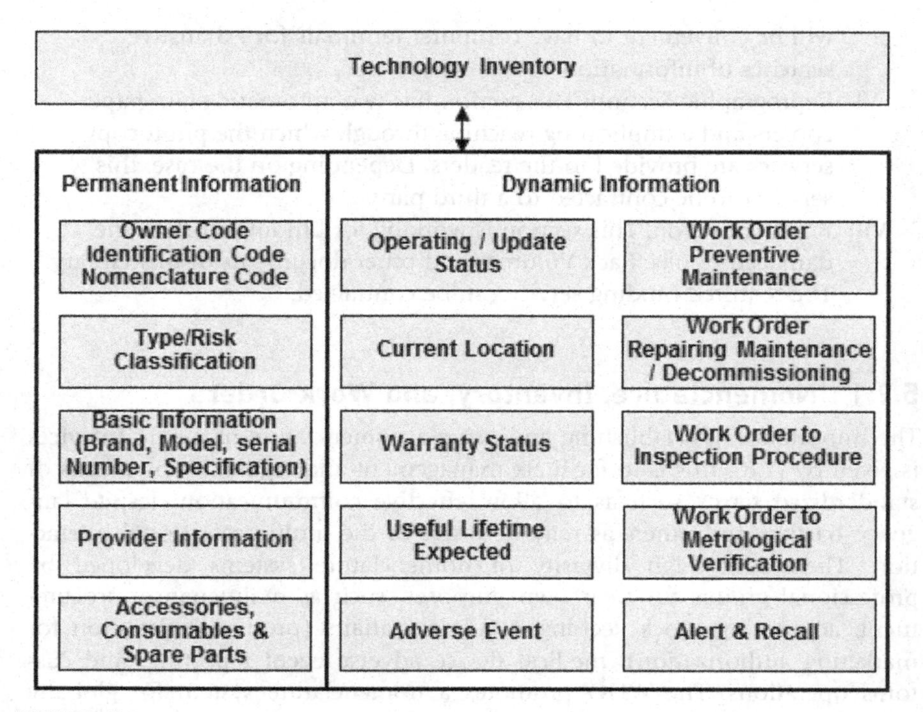

FIGURE 5.2
Information components for inventory in an integrated healthcare technology management system.

dynamic, and the warranty status changes over time. From the point of view of computer resources there is no limit on the extension and handling of entries for the database, the important thing is how we want to visualize the information in the computer terminal and carry out the analysis of information following trends of the indicators for determining efficiency, safety, and costs associated with each technological items. Fig. 5.2 shows a summary of information that may contain a technology inventory. Of course, depending on the type of technology you can have some variations.

5.2.2 Information System for Health Technology Management

An information system is a software system to capture, transmit, store, retrieve, manipulate, or display information, thereby supporting people, organizations, or other software systems. It is about helping people to find the information they seek, whether internal institutional information or external information without limit on all sources of information. There are many software for managing medical equipment (HEMS-EQ2, Computerized

Maintenance Management System (CMMS), KeyTone Technologies, Asset360 Equipment Management Program, Vizbee Healthcare, AeroScout Healthcare Visibility Suite, MantHosp, etc.). The WHO promotes the CMMS. The information included in the CMMS [10] varies depending on the individual situation but always includes the medical equipment inventory and typically includes information such as service history, preventive maintenance procedures, equipment and performance indicators, and costing information. However, an information system for the integrated HTM is still to be developed. We believe that software for holistic management of technology is required, for managing all processes of HTM, including the management of technological assets, and that they also consider all types of technology (see Chapter 1: Healthcare Technology Management (HTM) & Healthcare Technology Assessment (HTA)), as well as specific applications for the technology management corresponding to each clinical services. This is an interesting challenge for software design and to meet the requirements of an information system for the branch of technology in hospitals.

5.3 OPERATIVE MANAGEMENT AND INSPECTION PROCEDURES FOR FUNCTIONAL TECHNOLOGY

The strategy for having medical teams operating in a hospital is posed by the WHO; it consists of three components: corrective maintenance, preventive maintenance, and inspections [11]. It is known that in many countries inspections are not implemented and they are not carried out. Likewise, there are still problems in the implementation of preventive maintenance due to the lack of clear policies, procedures, and resources to carry them out. However, within the framework of the holistic approach to technology management, functional technology is sought in the hospital, i.e., operational technology but with efficiency and security in its use and application, with cost control. We reiterate that we do not refer only to medical equipment, but that clinical services require the whole technological environment to be functional, i.e., all clinical technologies (clinical procedures, medical equipment, and medicines), support technologies (infrastructure, systems Information, hospital equipment, and organization), and technologies for community health. In this sense, some technologies require the triad: corrective maintenance, preventive maintenance, and inspections, but others require something different, medicines require good storage conditions to be optimal at the time of application, soft technologies require updates or reconfigurations, and many others require functional testing, evaluations, and simulations. For corrective maintenance and preventive maintenance there exist a lot of experience, methods, and information in an individual way, but it is a question of gathering them, integrating them in order to manage them under a management system with systematized processes for rapid response and high repetition. The design of these processes and their systematization are a creative work for clinical engineers and biomedical engineers, as well as their application and improvement according to the results obtained.

Corrective maintenance is an unscheduled task, but if all your events are properly recorded, you will be able to analyze this information and provide planning actions and budget for the next period. A first decision to make is the determination of which items correspond to corrective maintenance and which to preventive maintenance. The methods proposed by WHO [10] and others such as Mike Capuano [12] and risk-based maintenance are useful for this decision. A second decision is the scheduling of preventive maintenance and inspections, e.g., it is necessary to determine the frequency of these interventions. It is remembered here that these are hundreds and perhaps thousands of items, so it is necessary to manage a consolidated method. On the other hand, one of the works of high relevance is the inspections, often confused with the preventive maintenance, and as said, they still need to be implemented in many hospitals. Under the holistic approach of technology management, it is necessary to expand inspections to observe all types of healthcare technology. The inspections consist of observing the technology operating in its environment, in its place of application, and it is recorded as a check list with the conditions of the observed aspects, for example

- to observe the management of technology by the users,
- to inspect the energy system conditions,
- to inspect configurations according to the clinical case,
- to inspect the application of clinical procedures, is it as effective as they say?
- to observe the operation of the processes to determine the improvements,
- to inspect the infrastructure and hospital equipment, ventilation, and air conditioning,
- to inspect the organization into the clinical services,
- to verify the functioning of the software,
- to inspect the procedures of metrology and its validity,
- to inspect the communication and connectivity,
- to observe the compatibility of the clinical cases with the equipment and the infrastructure,
- to verify the environmental conditions,
- to observe the cleaning processes,
- to take many other actions in order to ensure the proper operation of the technology.

That is, inspections are actions to ensure quality rather than a maintenance action. The most important thing is the information collected and the actions taken from it. Naturally, inspections bring engineering personnel closer to health professionals. In some hospitals, the presence of engineering staff at the time of change of health professionals' guards is an outstanding activity to observe and adapt the technology environment to the clinical requirements of the moment. Once again this approach makes the work of a clinical engineer or a biomedical engineer in hospitals more attractive and interesting.

We will then review some of the most relevant actions in technological assets management in order to achieve a functional technological environment in hospitals.

5.3.1 Installation and Functional Testing

In Guide 3: how to search and commission your healthcare technology [13], it is very well established that "Many poor installation and commissioning practices are due to poor communication and coordination between the various responsible departments and the different types of staff. Common problems include"

- equipment arriving on site unannounced;
- contractors arriving to work in the health facility without giving prior warning;
- contractors carrying out work in the health facility without consulting the users and in-house technical staff;
- health service staff left wondering what is happening and unable to plan their work.

This problem can be extended to all types of technology existing in a clinical service. Who is responsible for the technology in a hospital? Medical care personnel are involved but they would not be responsible since their role is another, and the administrative staff is also not responsible since the professional profile for the installation and functional tests corresponds to a professional profile of clinical engineering or biomedical engineering. Our approach is to have a branch in charge of technology that is effectively responsible for the technology in all its details and processes. In the following, Table 5.2 shows examples of the different actions to be taken for this first stage of technological assets management. Naturally, the details can be further developed according to the type of hospital and local environment, but the idea is to provide this holistic approach to a comprehensive HTM.

The commissioning process of health technology begins with safety tests. In the case of hard technologies, the physical aspects such as electrical or radiation are highlighted, followed by adjustments and metrological verification, functional tests, and recording results, which can generate an approval

Table 5.2 Examples of Actions to Assembly, Installation, and Initial Training According to Type of Health Technology and Clinical Service

Technologies for Healthcare	Clinical Service or All Hospital					
Individual Health Services						
a. Clinical technologies						
Clinical procedures	Ensure the work site is ready	Identify staff to learn from commissioning and monitor the work	Liaise with clinical staff installation team	Provide the necessary procedure	Ensure compliance with standards and regulations	Provide necessary working space
Medical devices	Ensure the work site is ready	Identify staff to learn from installation and commissioning and monitor the work	Liaise with the in-house/ visiting installation team	Provide the necessary inputs (support to installation team, materials, safety testing instrument, etc.)	Ensure compliance with standards and regulations	Provide necessary working space
Pharmaceuticals & others consumable products	Ensure the storage site is ready	Identify staff to learn from commissioning and monitor the work	Liaise with the in-house/ visiting reception team	Provide the necessary materials	Ensure compliance with standards and regulations	Provide necessary working space
b. Support technologies						
Organization & management system	Ensure the influence area in the organization is ready	Identify staff to learn from commissioning and monitor the work	Liaise with clinical & administrative staff installation team	Provide the necessary inputs	Ensure compliance with policies and regulations	Provide necessary working space in the organization
Infrastructure & hospital equipment	Ensure that the work site is ready	Identify staff to learn from installation and commissioning and monitor the work	Liaise with the in-house/ visiting installation team	Provide the necessary inputs	Ensure compliance with standards and regulations	Provide necessary working space
Information & communication technologies	Ensure the work site is ready	Identify staff to learn from installation and commissioning and monitor the work	Liaise with the in-house/ visiting installation team	Provide the necessary inputs	Ensure compliance with standards and regulations	Provide necessary working space

Continued

Table 5.2 Examples of Actions to Assembly, Installation, and Initial Training According to Type of Health Technology and Clinical Service *Continued*

Technologies for Healthcare	Clinical Service or All Hospital					
Community Health Services						
Protection technologies	Ensure the storage site is ready	Identify staff to learn from commissioning and monitor the work	Liaise with clinical & administrative staff installation team	Provide the necessary inputs	Ensure compliance with standards and regulations	Provide necessary working space
Prevention technologies	Ensure the storage site is ready	Identify staff to learn from commissioning and monitor the work	Liaise with clinical & administrative staff installation team	Provide the necessary inputs	Ensure compliance with standards and regulations	Provide necessary working space
Promotion technologies	Ensure the influence area is ready	Identify staff to learn from commissioning and monitor the work	Liaise with clinical & administrative staff installation team	Provide the necessary inputs	Ensure compliance with standards and regulations	Provide necessary working space
Environmental Health	Ensure the work site is ready	Identify staff to learn from commissioning and monitor the work	Liaise with the in-house/ visiting installation team	Provide the necessary inputs	Ensure compliance with standards and regulations	Provide necessary working space

Modified from Guide 3: how to procure and commission your healthcare technology [13].

certificate. In the case of soft technologies, we begin by reviewing the compliance with regulatory aspects or institutional policies, then, the preparatory actions and the functional tests are performed to finally record the results and generate an approval certificate. In all cases, initial training should be considered, which involves preparing the contents, study materials, examples or simulations, and testing or experimentation.

5.3.2 Requirements for Supplies, Materials, Accessories, Consumables, and Spare Parts

The vast majority of health technologies need to be supplemented with supplies and materials to achieve proper functioning; additionally, accessories, consumables, and spare parts are often more linked to the operation of hard technologies, medical devices, hospital equipment, power systems, etc. There are definitions for accessories and consumables such as Guide 4: how to operate your healthcare technology effectively and safely [14], where accessories are parts that connect to the equipment (breathing circuit, ECG conductors, and transducers), They facilitate their use (foot switches and computer mouse) and adapt their operation (adapters for adults or children and lenses). And consumables are those items that are used daily for the operation of the equipment (gel, gloves, gauze, needles and syringes, and disposable electrodes) and will be needed over the life of the equipment. These definitions do not always permit the distinction of one with another, both can be of single use or disposable and both can be also reusable, reason why the discussion continues. The important thing is that purchase and stock control must be guaranteed and this will depend on the lifetime of each of them.

For an equipment to last the expected lifetime, some parts of the equipment need to be replaced. These spare parts need to be available all this time. Maintenance materials (oil, grease, washers, fuses, etc.) must also be available for repair and preventive maintenance work. Guide 5: how to organize the maintenance of your healthcare technology [15] states that most of the repair workload and maintenance consist of simple tasks and these require spare parts, thus 80% of all required spare parts represent less than 20% of spare parts total cost, then 20% of the repair and maintenance tasks consume 80% of the spare parts budget due to complex and expensive equipment. The availability of an up-to-date and complete inventory will allow for a minimum stock of spare parts and maintenance materials. This should include not only the equipment, but the accessories, spare parts, maintenance materials, and consumables of the equipment. In the holistic approach to technology management, the planning, inventorying, and proper recording of work orders such as maintenance and inspections allow for forecasting and reporting of requirements for supplies, materials, accessories, consumables, and spare parts for all types of technology in a clinical service, so each item should be registered in the information system linked to the corresponding technology. Again, this job classification, organization, and process design will be of interest to clinical engineers and biomedical engineers in hospitals.

5.3.3 Functional Inspection Procedures & Preventive and Corrective Maintenance

To ensure functional technology in hospitals it is necessary to do much more than the maintenance of medical equipment. In other words, the operation of medical equipment is not enough knowing that the technological environment is not only medical equipment, but a set of soft and hard technology types interacting in a coordinated way, which together makes it useful for health services. Under the holistic approach of HTM, corrective repair or maintenance must bring the different technological components back to operation. It can be easily understood that hard technologies need to be repaired, but how to "repair" a soft technology? a process? an organization? When the soft technologies do not work, it is necessary to intervene to make them work again. In the same sense, preventive maintenance is a periodic intervention procedure to minimize the risk of failure and ensure continuous operation of technology, hard and soft technologies. In both cases, an intervention is also possible. But, neither corrective maintenance nor preventive maintenance can ensure the operation of all the technologies in a clinical service, because both are individual rather than joint interventions. In addition, corrective maintenance and preventative maintenance also do not provide workplace information, in situ, so we can ask how we ensure the functionality of the technology during the time of use? We believe that these and many other premises allow us to justify the need for inspections, which consists of periodic verification of the performance and safety of the individual and collective technologies in the workplace, such as clinical services. The functional inspection of technologies, not just medical equipment, should be a new challenge to be developed to guarantee the quality of health services based on the quality of the technological environment.

Corrective maintenance, preventive maintenance, and inspection of medical equipment are abundant information that can be consulted [11,14,15] and many others including templates, models, and cases. However, in extending the concepts to the different types of technology, it remains for the clinical engineers and biomedical engineers a detailed design work, diverse and therefore interesting, that requires specialization and therefore of multidisciplinary teamwork for the development of standard operating procedures (SOPs) for each component of each type of technology. Of course, prioritization should be considered in order to carry out work that is viable in practical and economic terms.

5.3.4 Management of Contracts and Attention to Suppliers

In the work of making functional the hospital technology, not everything can be done by the hospital's own staff, but it is necessary to have capacity to manage, supervise, and evaluate the work performed by third parties. If we consider the holistic approach of technology management, not only is it necessary to contract the maintenance of medical equipment, there are different requirements for hiring companies or external personnel in order to carry out studies, consultancies, formulate projects, develop strategic plans, design of software, reports of technology assessment, process design, clinical trials, etc. Also, private commercial companies are usually partners of the hospital and can offer a range of services that complement the capacity of the hospital.

Similarly, public sector entities or other hospitals can provide support in specific and specialized cases. A factor that induces third party contracts is the complexity of the technology, because to ensure its operation requires trained personnel and expensive test equipment, which is costly for the hospital. In the traditional model of medical equipment management, contracts are usually punctual or specific and end up with putting the equipment in operation and then, the repaired equipment is returned to the clinical service and there ends the work of the engineering department. In the holistic approach of technology management, engineering staff has the responsibility of integrating the equipment into the clinical service and verifying the functionality of the technological environment as a whole. Likewise, for negotiation processes with third parties, it is clear that the hospital must have an appropriate counterpart that values the proposals and finds the correct correlation between the final product and the costs. To do this, the hospital must have someone responsible for its technological environment and not just medical equipment. In the traditional model of medical equipment management, a person responsible for the technological environment is not always identified, which can generate poor counterproductive contracts for the hospital.

In an integrated and comprehensive health technology management system, there is a better capacity to manage contracts through Public–Private Partnerships (PPP). This modality implies that a private entity contracts with the state to build infrastructure and provides services that are usually provided by the state, fulfilling the following conditions: (1) the state always owning the infrastructure, even if the private entity is in possession while the project lasts; (2) the construction, operation, and maintenance are integrated into the project design, which obliges to meet standards for the benefit of patients and the owner state; and (3) a contract regulates the obligations of the parties for a defined period, as well as the state continues to fulfill its regulatory role in infrastructure specifications, risk allocations, environmental demands, guarantees, insurance, quality of services, etc. In other words, a PPP contract allows a balance between the interests of the state that is to promote health services with quality, and interest of the private entity that is to have profitability, all without cost to the user [16]. PPPs typically cover three types of services: (1) the "white coat," include clinical services that are rarely part of the obligations, because these services are provided by the state's own physicians; (2) the "green coat," associated with the operation and maintenance of intermediary services such as laboratory, diagnostic imaging, and sterilization; and (3) the "gray coat," associated with complementary services such as laundry, surveillance, waste management, security, and maintenance of equipment.

PPP's will play a critical role in improving health systems, especially in developing countries, bringing together the best characteristics of the public and private sectors in order to increase efficiency, quality, innovation, and the impact of health on the population. In this scenario the holistic approach to technology management is crucial to guarantee the expected functionality of the technological environment, especially when considering the "green coat" and the "gray coat," which are precisely the areas of intervention of PPP's.

Returning to contract management, let us remember that this is the process by which people or companies sell services to the hospital, will be remunerated but will also be supervised [15]. Thus, it is suggested that these persons or companies be registered as clients of the hospital fulfilling requirements aligned with institutional objectives and policies. The registration will facilitate the verification of the profile of the person or company, payments will be simpler, you will usually get quick response from registered individuals or companies and corruption or inappropriate payments to hospital staff members is less likely.

5.3.5 Evaluation and Execution for Asset Retirement

It is possible to estimate the useful life of medical equipment and hospital equipment, it is also possible to estimate them in the case of medicines, some prevention technologies, protection, but in some others such as organization, clinical procedures, software, and hospital infrastructure. Require a different analysis and are subject not only to the life cycle in the market, but also to government policies, technological trends in competition, or new benefits that can mean cost savings and better capabilities. In the holistic approach to technology management, processes of decommissioning, dismantling, elimination, or replacement of all types of technology are considered, but with a broader and therefore more inclusive perspective to predict impacts on health services. On hard technologies we could say that they depreciate over time due to daily use, how it has been managed, maintained, and cleaned, the impact of the environment such as temperature, humidity, and pollution. With respect to soft technologies, it could be said that they are de-updated, lose their validity, or the environment changes to a scenario with different characteristics, which implies to bring about improvements or changes. In this sense the research and monitoring of the evolution of the technologies help a lot to identify these changes and propose the corresponding improvements. It is observed that many hospitals maintain the same processes and practices throughout the years and at the same time they have one or another equipment of last generation, which in an approach of integrated management of healthcare technology would not be given, since first, by having the formal responsibility for technology and then because the holistic perspective allows visualizing the improvements and technological development of the whole.

Policies for the replacement of technology are given in abundant information. In Guide 4: how to operate your healthcare technology effectively and safely [14] establish some criteria for the replacement of hard technologies: It is worn or damaged beyond repair, it is not reliable or insafe, clinically obsolete, spare parts are no longer available, it is not economical to repair, there is a demonstrated clinical or operational need. It should be noted that replacements are not justified just because the device is old or does not like staff or there are new models on the market.

5.3.6 Medical Technology Support

In the traditional models of medical equipment management, oriented solely to maintenance, the responsibility for the technological development of a hospital is ignored. The traditional engineering department is usually receptive of the needs for repair and maintenance of medical equipment. However, in the model of integrated system of HTM or technology branch, it is in charge of the technological development of the hospital, the attitude is propositive to incorporate validated improvements and innovations in technology for the best care of health services, and for this purpose we have methods such as healthcare technology assessment (HTA), clinical trials, meta-analysis, and evaluation of technological trends.

The following is a summary list of technologies that clinical engineers and biomedical engineers should be considering to incorporate or improve in hospitals [17], which involves the design of clinical services that functionally contain these technologies, specifying the requirements of infrastructure, energy systems, information and communication systems, medical equipment, organization, clinical procedures, pharmacology, medical materials, and environmental conditions:

- Organ and tissue donation system
- home telemedicine
- primary care telemedicine
- adequacy of ambulatory care services to telemedicine services
- breast milk bank service
- lab-in-a-backpack: point of care screening/diagnostic
- microbial water testing kit
- mobile phone image transmission for diagnosis
- portable cell sorting and counting device
- portable system for pre-cancer screening at point of care
- portable telemedicine unit
- portable transcutaneous hemoglobin meter
- technology based on genetic engineering, stem cells, and tissue regeneration
- disposable devices reprocessing unit

- shared electronic health records system
- occupational health service
- fetal heart rate monitor by mobile phone
- portable field hospital in disaster area
- ambulances for rural areas

5.4 MANAGING RISK FOR FUNCTIONAL TECHNOLOGY IN CLINICAL SERVICES

Dangers are inherent in technology, in other words, technologies imply a risk factor that must be managed in order to control it and reduce it to a reasonable minimum, just as technology is managed to maximize its benefits with patients. In short, HTM involves working to reduce risks and maximize the benefits of technology. In the traditional model of maintenance or engineering departments, emphasis is placed on the benefits based on operative equipment, and risk management is a much more recent fact that is not yet well established in the organization of hospitals.

In addition, we should consider the approach given by [18] "To Err Is Human," which reports that the incidence of adverse events in New York is between 2.9% and 3.7% of patients hospitalized in 1999, many of which are not due to the physician in his/her self, but to the health system, that is, its procedures and technology. Not all countries have studies in this regard, estimates are scarce except for some reports such as [19]. However, there are thousands of affected patients who should not be harmed. In addition to the harm done to patients, another aspect of this study shows the high cost to affected individuals, families, the health system, and the country as a whole. Loss of economic income, loss of production, disability, and costs of health services are generated due to the additional services required. The decentralized and fragmented nature of the healthcare delivery system (also called "nonsystem") also contributes to unsafe conditions for patients, and serves as an impediment to efforts to improve safety [18]. The authors declare the need to break with this circle of inaction because this status-quo is not acceptable and can't be tolerated anymore. We believe that an integrated HTM system, a branch of the organization responsible for all types of technology used in a hospital, will be able to more appropriately address all aspects of technology that end up hurting patients, take responsibility for the whole technological environment, and therefore has obligation to learn from their own mistakes [20]. And at the same time, control the costs associated with these events or, in other words, reduce the waste of resources due to resolving events that should not have happened.

The dangers and risks related to medical equipment are widely known and in fact there is abundant information published. Infection control also has abundant information and practical guides. On the other hand on solid waste management and disaster mitigation in hospitals, there are guidelines, rules, and regulations. However, we observe in many hospitals that these issues are handled by committees or internal units, which may be the right thing to start a new line of work, but these do not develop to the point of being considered a functional responsibility, that is, that hospitals have the due processes, functions, methods, trained personnel, and budget for proper management. In the traditional model of hospital organization, engineering departments are focused on the operability of medical equipment and are likely to participate in safety committees, but often do not incorporate functions or responsibility in the subject, therefore, they usually do not have human or material resources and their intervention ends up being a collaboration. It is true that the engineering department can acquire some functional responsibilities regarding the risk management, in fact happens in some hospitals, however, they are not usually preponderant for the organization of the hospital and many actions identified and raised are left aside. Under the holistic approach to HTM, the branch of technology in the organization of the hospital is responsible for the technology in the hospital and must functionally, within its own organization, assume the processes related to risk management. Fig. 5.3 summarizes the main processes to be considered for healthcare technology risk

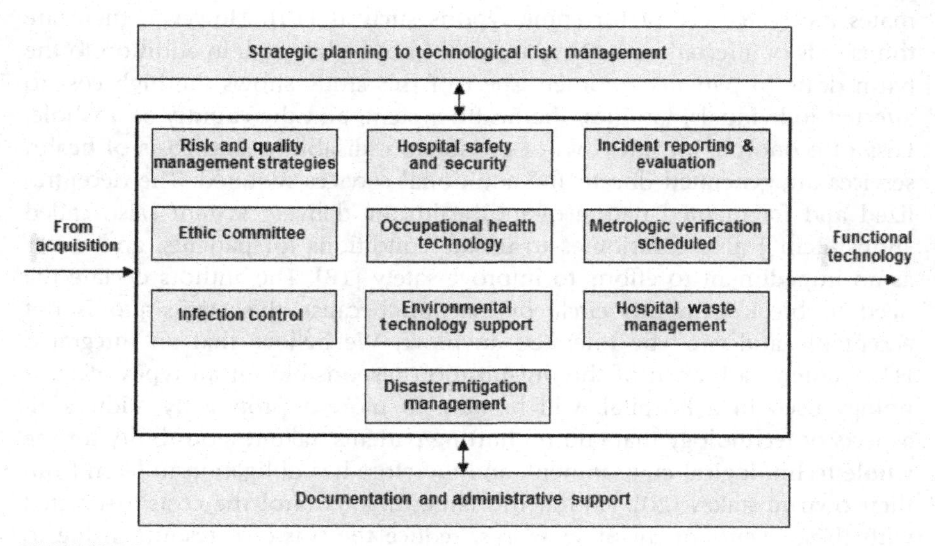

FIGURE 5.3
Examples of process map to technological risk management in a hospital.

management. Of course, each hospital will have to adapt these processes to its level of complexity and resolution.

5.4.1 Techno Surveillance: Technology Risk Management in Clinical Services

It is very true that safety in the use of medical devices is of vital importance, in fact there is plenty of information about it. Guide 4: how to operate your healthcare technology effectively and safely [14] describes safety aspects to consider, including specific hazards in the operation of medical equipment, infections control related of equipment and their decontamination, handling of waste related to medical equipment and other systems such as gas, electricity, and radiation which require attention when operating medical equipment. It also describes electrical and mechanical tests to ensure the reliability of medical equipment. We believe that these important tasks are not always sufficiently established in the organization and processes of the hospital, in addition they are very focused on the equipment itself when the center is the patient and the users, and lastly, these tasks should extend to all kinds of technology in order to have a safe technological environment for patients. In the holistic approach to technology management, these and other premises about safety can be assumed by a functionally responsible unit through its processes and human and material resources. At this point, the job is to monitor how technology is being used, verify compliance with regulations, standards, and institutional security premises. This safety oversight work is complemented by performance inspections of the technologies described in Section 5.3, so more comprehensive inspections could be conducted to ensure the functionality of clinical services, considering clinical effectiveness, use of technology, safety, cost control, and taking action to resolve the events found. In [14] more common examples of hazards during the operation of medical equipment are listed, but what would be the hazards of infrastructure? the electrical system, ventilation system, gas management, high ambient temperatures, or low temperatures, the danger of reaching the dew point before changes in humidity and temperature, etc. But we must also identify dangers in soft technologies, information sent by data networks, the confidentiality of information, its protection over time, the danger of not having a process or procedure for an important event, for example in case of emergencies or disasters.

In addition to complete inspections, including safety surveillance, Hazard Identification and Risk Assessment (HIRA) should be performed for each clinical service, considering all types of technology. This is a new and interesting task to perform for clinical engineers and biomedical engineers. A specialized professional profile with design and analysis capabilities is required.

HIRA is the process for defining and describing hazards, in this case technology, by characterizing their probability, frequency, and severity, then assessing their adverse consequences including potential losses and damages, and finally determining the actions of control, safeguards and strategies to mitigate risks. The HIRA process involves four different steps: (1) Identification and study of hazards/risks. (2) Carry out the risk assessment for each identified hazard to determine the probability of occurrence and the consequences in people. (3) Establish a program of priorities to be addressed using a risk assessment matrix. (4) Develop a specific plan for prioritized hazards. HIRA processes can be complemented with those developed for occupational health in employee protection, but it is important here to identify the dangers of the technology that may impact on patients and users. There is a wealth of information on HIRA and OHSAS 18001 for use in industry and services, but it can be said that its application to hospital technology is still ongoing and it is advisable to promote this practice for the benefit of patients.

5.4.2 Techno Surveillance: Technology Adverse Event Management—Registration and Resolution

The purpose of reporting adverse events is to learn from mistakes and take actions so that they do not happen again. Very often, neither healthcare providers nor health organizations warn others about their mistakes, nor do they share what they have learned when an investigation has been conducted. As a consequence, the same errors crawl repeatedly and patients continue to be harmed by preventable errors [20]. The report of adverse events is the first step to solve these problems, and the second is logically the analysis of these for taking actions. The report is crucial and should be done by health professionals, including engineers, the hospital organization itself, or by organizations with broad regional or national roots. In some places patients themselves may report adverse events, although it would require completing the hospital's own information. It is about identifying hazards and risks, and providing information on the weak points of the health system. This can help direct improvement efforts and changes in the health system in order to reduce harm to future patients. From the ethical point of view, it is understood that the adverse events report is not the method to manage the risks, it is assumed that the hospital already has an action plan in place for risk management and has implemented processes to perform them, in change, the report of adverse events is to identify failures in an existing management system, so it is considered a way to improve the quality of health services. In the holistic approach to health technology management, the adverse events report includes those related to health technology, naturally considering all types of technology, hard or soft, clinical or supportive or community. Beyond that only medical equipment, this approach opens up a much more interesting field of analyzing the causes of adverse events and thus a new perspective of professional work by clinical engineers and biomedical engineers in hospitals. It should be considered that the technological causes of adverse events are only one of the areas of study of health services (branch of technology), the other are administrative aspects (administrative branch) and the medical or proper aspects of the services attention (medical branch).

Reports of adverse events are performed in fairly structured formats but always with at least one space for a narrative description of the event. Then a classification of the error is performed, for which there are different systems called taxonomies. The WHO—World Alliance for Patient Safety promotes a taxonomy [20] aimed at developing an international system. Taxonomies usually fall into three categories: classification by event, by risk, or by cause. The classification system to be used must be designed according to the objectives to be achieved, the type of information that will be available and the sources of information, which will surely have a cost when implementing it. The next process is the report analysis. These should at least allow the identification of a new and unexpected danger. The analysis must also generate a summary and a description to determine the occurrence frequency of the event, as well as trends in time to identify new problems. The use of statistical methods will be necessary. A series of events that suddenly arise suggest the need for research. The trends analysis allows to find correlation with other variables. Finally a risk analysis is done and the causes are identified, which will allow taking the pertinent actions. In order to carry out this system of adverse event reporting and analysis, clear technology policies are required and an integrated technology management system, formally accepted in the hospital organization, can be a great help in consolidating event reports adverse on the health system.

5.4.3 Technology for Occupational Health

Safety and health in the workplace of hospitals is another area that requires technology and therefore involves the integrated HTM system, and in particular the risks management due to technology. It may be considered an additional health service that hospitals have been implementing in recent years. Usually the engineering department in the traditional model is involved very little, however, we anticipate that a holistic approach to technology management will consider it more appropriately. It is about avoiding occupational diseases and accidents at work, but also avoiding the costs derived from them. For occupational health in hospitals, preventive and protective measures should be taken, so that they are implemented in the following order of priority [21]:

- Elimination of the hazard/risk
- Control of the hazard/risk at source, through the use of engineering controls or organizational measures
- Minimization of the hazard/risk by the design of safe work systems, which include administrative control measures
- Where residual hazards/risks can't be controlled by collective measures, provision by the employer of appropriate personal protective equipment, including clothing, at no cost, and implementation of measures to ensure its use and maintenance

The most significant risks for a hospital are (1) biological agents, (2) musculoskeletal disorders, (3) psychosocial disorders, and (4) chemical agents. Specific risks that should be addressed are, for example

- the handling of blood and blood products, including the handling of needles and other sharp objects;
- exposure to chemical agents/hazardous substances, including cleaning agents and disinfectants;
- time pressure, high workload, and interpersonal conflicts;
- bullying or violence at the workplace;
- shift, weekend, and night work;
- manual patient handling, lifting, pushing, and pulling of weights;
- the ergonomic design of workplaces.

We consider that an integrated HTM system, which aims to make technology functional in a hospital, taking into account all types of technology, including clinical procedures, medical equipment, and assistive technologies, and these technologies will provide a better contribution to the hospital's occupational health services due to its ability to design clinical services, analysis, and research. This activity is of great interest to clinical engineers and biomedical engineers.

5.4.4 Metrological Verification Management

Many countries show weakness in their metrological laboratory systems or networks for health technology, which discourages certified calibration processes and metrological verification within hospitals. The traceability of the patterns is usually a serious problem when determining the measurement errors of medical devices. As it is known the repeatability of the measurements is defined by the precision, which in turn is defined by the manufacturer through its designs and components used, to some extent represents the quality of the medical device, however, another aspect of utmost importance is the accuracy of the measurement, which is the determination of the measurement error between the value obtained by the equipment under test and the actual value given by an equipment calibrated (equipment with known traceability). Precision and accuracy are the central points of biomedical metrology as it defines with what certainty is taken from the data given by the medical devices and then, they are used for all kinds of clinical procedures, diagnoses, treatments, and therapies. Estimates of false positives and false negatives have been reported due to measurement errors of some medical equipment [22]. In the traditional model of maintenance of medical equipment the objective is the operation of these, therefore, usually overlook some aspects of quality, in this case the quality of the measurements. On the other hand there is some information on biomedical metrology but also requires a great work to implement and adapt a system of metrology for hospitals, however, experience shows that it has been very difficult for a hospital to implement this type of programs.

In the holistic approach of HTM, since the objective is the functionality of the technology, with clinical effectiveness, safety, and efficiency in the use of technology, it becomes evident the need to guarantee minimum measurement errors and raises the metrological verification processes of medical devices and the traceability of local pattern equipment. This capability involves having qualified laboratories [23,24,25], certified pattern simulators, measuring instruments, a metrological data recording system, design of SOP's for metrological verification by each type of medical device and count specialized professionals. These services can be contracted to third parties or have a mixed capacity, for example with the own resources to meet the requirements for devices of low and medium complexity and to contract services to third parties for those of high complexity. There are no complete studies, but the cost of endures false positive and false negative diagnoses and treatments will be much higher than the cost of implementing a biomedical metrology program. Of course all these initiatives will be possible if there are clear policies in health technology, such as to enable a series of consequences such as have a traceability system of standards for health technology, strengthen the work of regulatory entities for compliance of biomedical metrology programs in hospitals, support hospitals for the implementation of their metrological verification capability for medical devices as a function of an effective improvement of the quality in health services for the benefit of patients. Within the framework of an integrated HTM system or technology branch in hospitals, the professional development of clinical engineers and biomedical engineers has a better perspective by including metrology processes, as it involves process design, research for improvements or innovation, as well as the need to systematize processes given the amount of medical equipment that a hospital usually has.

5.4.5 Infection Control & Hospital Waste & Disaster Mitigation Management

A typical engineering department is in charge of medical devices. The WHO medical device maintenance program [11] provides guidelines for engineering personnel not to be contaminated when repairing or intervening in medical equipment, but engineering staff involvement in the infection control committee is minimal and tangential, and is that in the traditional model of maintenance of medical equipment the profile of engineering professionals does not correspond to other interventions in technology. In the holistic approach to HTM, the professional profile of engineering corresponds to that of clinical engineering or biomedical engineering, with training in physiology and technology in health, so we foresee a much more active participation in the control of hospital infections or hospital waste management, and even in hospital disaster mitigation programs.

The important components of the infection control program are [26]

- basic measures for infection control, i.e., standard and additional precautions;
- education and training of healthcare workers;
- protection of healthcare workers, e.g., immunization;
- identification of hazards and minimizing risks;
- routine practices essential to infection control such as aseptic techniques, use of single use devices, reprocessing of instruments and equipment, antibiotic usage, management of blood/body fluid exposure, handling and use of blood and blood products, and sound management of medical waste;
- effective work practices and procedures, such as environmental management practices including management of hospital/clinical waste, support services (e.g., food, linen), and use of therapeutic devices;
- surveillance;
- incident monitoring;
- outbreak investigation;
- infection control in specific situations; and
- research.

An integrated HTM model, which considers all types of technology (see Chapter 1: Healthcare Technology Management (HTM) & Healthcare Technology Assessment (HTA)), hard technologies (such as infrastructure, hospital equipment, medical devices, and energy systems) and soft technologies (such as clinical procedures, human resources based on clinical engineering, biomedical engineering, and medical physics), has a better ability to effectively support the hospital Infection Control Committee and to support the implementation of all components of the program given its capacity for design, analysis, supervision, and research in health technology.

On the other hand, hospital waste management establishes a series of actions such as the characterization of waste in the hospital, evaluation of the risks associated with hospital waste, compliance with regulations and standards, planning of hospital waste management, the processes of minimization, re-use and recycling, segregation, storage and transportation of waste, methods of treatment and disposal, treatment of wastewater, economics in hospital waste management, occupational health practices, hygiene and infection control for program workers, and training [27]. However, hospitals often have complications in the implementation of these directives, because they are not tasks exactly for the administrative branch and not for the medical branch, but they are absolutely necessary. In the same sense, a branch of technology, with an integrated HTM system, can be very useful for directing hospital waste management efficiently and safely for the benefit of the hospital, in addition to opening a field of interesting development for hospital engineers, clinical engineers, biomedical engineers, and others professionals. In the same sense, the mitigation of disasters in hospital facilities, whether caused by natural or induced disasters, is a vital activity. Hospitals must, to a certain extent, ensure their operation after a disaster, so it is necessary to carry out risk reduction in hospitals, structural vulnerability analysis and nonstructural vulnerability analysis, internal emergency care and external emergency care [28]. The holistic approach to technology management allows for the incorporation of these functions in a structured way in the organization of hospitals, which will create an excellent capacity for hospital resolution expected by the attached population.

References

[1] Jin Z. Global technological change: from hard technology to soft technology. 2nd edition. Chicago: Intellect Publisher; 2011.

[2] Coe RM. Sociología de la medicina. Madrid: Alianza Editorial; 1973. p. 271–2.

[3] Frisch P. What is an intelligent hospital? A place where technology and design converge to enhance patient care. IEEE-EMBS Pulse, November/December 11; 2014.

[4] Haux R. Health information systems—past, present, future. Int J Med Informatics 2006;75:268–81.

[5] Tomasi E, et al. Health information technology in primary health care in developing countries: a literature review. Bull World Health Organ November 2004;82:867–74.

[6] Sharma CK, Singh K. Library management, vol. 1. New Delhi: Atlantic Publishers and Distributors; 2005.

[7] GMDN User Guide Version. A comprehensive guide to the Global Medical Device Nomenclature. Oxford: GMDN Agency Ltd; 2010.

[8] WHO. Stakeholders' informal consultation on nomenclature for medical devices. Department of essential health technologies. Geneva, Switzerland: WHO Headquarters; 23–24 March 2011.

[9] Introduction to medical equipment inventory management. WHO medical device technical series. Geneva: World Health Organization; 2011.

[10] Computerized maintenance management system. WHO medical device technical series. World Health Organization; 2011.

[11] Medical equipment maintenance programme overview. WHO medical device technical series. World Health Organization; 2011.

[12] Capuano M, Koritko S. Risk oriented maintenance. Biomedical Instrumentation and Technology 1996;25–37 January/February

[13] Guide 3: how to procure and commission your healthcare technology; Ziken International. Series Editor: CarolineTemple-Bird, UK. 'How to manage' series for healthcare technology. 2005.

[14] Guide 4: how to operate your healthcare technology effectively and safely; Ziken International. Series Editor: CarolineTemple-Bird, UK. 'How to manage' series for healthcare technology. 2005.

[15] Guide 5: how to organize the maintenance of your healthcare technology; Ziken International. Series Editor: CarolineTemple-Bird, UK. 'How to manage' series for healthcare technology. 2005.

[16] Mitchell M. An overview of public private partnerships in health. Boston: Harvard School of Public Health; 2000.

[17] WHO. Compendium of new and emerging health technologies. World Health Organization; 2011.

[18] Kohn LT, Corrigan JM, Donaldson MS, editors. To err is human—building a safer health system. committee on quality of health care in America. Washington, D.C: Institute of Medicine. National Academy Press; 1999.

[19] Nagel JH, Nagel N. Patient safety—a challenge for clinical engineering. Medicon 2007. In: IFMBE proceedings, vol. 16; 2007. p. 1043–6.

[20] WHO. Draft guidelines for ad verse event reporting and learning systems—from information to action. Geneva: World Alliance for Patient Safety. World Health Organization; 2005.

[21] European Commission. Occupational health and safety risks in the healthcare sector. Luxembourg: Publications Office of the European Union; 2011.

[22] Symposium celebrating 10th anniversary of the CIPM MRA October 8–9, 2009 and economic impact of metrology, Dr Franz Hengstberger—Member of the CIPM. CIPM: International Committee of Weights and Measures.

[23] International Organisation for Standardisation. ISO 13485:2003. Medical devices—Quality management systems—Requirements for regulatory purposes, Geneva, Switzerland, 2008.

[24] ISO/IEC 17025:2005. General requirements for the competence of testing and calibration laboratories. Geneva, Switzerland: ISO; 2005.

[25] IEC 60601 Series of technical standards for the safety and effectiveness of medical electrical equipment.

[26] WHO. Practical guidelines for infection control in health care facilities. Geneva: World Health Organization; 2004.

[27] WHO. Safe management of wastes from health-care activities, 2nd ed. World Health Organization; 2014.

[28] Principles of Disaster Mitigation in Health Facilities. Disaster mitigation series. emergency preparedness and disaster relief coordination program. Washington, D.C, Geneva: Pan American Health Organization. Regional Office of the World Health Organization; 2000.

Further Reading

Howell JD. Technology in the hospital: transforming patient care in the early twentieth century. Baltimore: Johns Hopkins. University Press; 1995. ISBN: 0-8018-5020-7.

Brown P, Sondalini M. Asset maintenance management—the path toward defect elimination. The evolution of maintenance practices. Lifetime Reliability Solutions HQ. ITA Training Associates; 2011.

Guide 1: how to organize a system of healthcare technology management.

Guide 2: how to plan and budget for your healthcare technology.

Guide 6: how to manage the finances of your healthcare technology management teams.

Siekmann L. Requirements for reference (calibration) laboratories in laboratory medicine. Clin Biochem Rev November 2007;28(4):149–54.

CHAPTER 6

Quality & Effectiveness Improvement in the Hospital: Achieving Sustained Outcomes

Science, Knowledge, Technology and Skills: Health Workforce making the Difference at the Health Organization

6.1 HEALTH TECHNOLOGY AND THE CONTEXT OF HEALTHCARE

According to Cook [1] technology can be an accelerant for delivering transformative change. The author remarks that the change expected is only possible if the investment in technology is planned, the staff that uses the technology has appropriate support and training, and the technology strategy actually delivers more productive healthcare at a lower cost. Aligned to this perspective Health Technology Management (HTM), Biomedical and Clinical Engineering are valuable and key knowledge areas aimed to improve the results obtained by the health staff.

Technology can transform the delivery of healthcare if organizations not only understand the opportunity to transform services, but are also able to realize those opportunities and actually deliver the changes (Nuffield Trust, Evidence for better Healthcare, 2017).

On the other side and according to World Health Organization (WHO), Health Technology Management & Planning, Biomedical Engineering (BE) and Clinical Engineering (CE) education and the training on related skills are required by healthcare staff in order to promote their achievement of affordable and sustainable quality healthcare delivery.

In the last few years scientific development supported by technology is gradually and sustainably changing the way to understand the processes of delivery, promotion, research, and the management of health around the world. Consequently, health organizations, particularly hospitals need skilled professionals to improve the making decisions and management aligned to factors such as (1) patient safety, (2) budget and costs, (3) quality, and (4) risk management.

Healthcare Technology Management Systems. DOI: http://dx.doi.org/10.1016/B978-0-12-811431-5.00006-0
© 2017 Elsevier Inc. All rights reserved.

6.1.1 Topics of Education for Healthcare Workforce

Quality and Effectiveness at the Healthcare organization require the Health Workforce be engaged to sustained education and training programs of HTM, BE, and CE. The following introduction presents the pertinence and value of this knowledge for the strategic objectives of the healthcare organization.

6.1.2 Health Technology Management—HTM

Discipline which focuses on ensuring access to appropriate medical devices, proper management, and use of medical equipment over its life cycle must be considered, beginning with understanding the needs of the country, region, community, or facility and ending with decommissioning. In between, the process consists of good procurement practices, appropriate donation solicitation and provision, logistics of delivery and installation, inventory management, maintenance, safe use and training, and measurement of clinical effectiveness. HTM is conducted alongside Health Technology assessment (HTA) and HT regulatory and performance compliance (WHO, Geneva, 2016).

HTM is the point of convergence of science, technology, and the market. Some of the competences related to HTM are to be able to manage the design, planning, development, commercialization, and regulatory compliance of diagnostic and therapeutic medical devices and the implementation, utilization, and assessment of healthcare technologies (American College of Clinical Engineering—ACCE, 2017). Other competences are related to the selection, implementation, usage effectiveness and management of healthcare technology equipment and systems, the budgeting for expansion or replacement of health technology systems.

Some key objectives of HTM are as follows:

1. Increase the availability of technology, through
   a. Proper maintenance
   b. Good purchase practices
2. Improve the patient care service
   a. Increase patient safety and cost management through standardization
   b. Reduce unnecessary hospitalization through Telehealth
3. Current and future cost management
   a. Reducing overall costs
   b. Reliable equipment replacement forecast
   c. Optimization of cash-flow
4. Financing and managing technology replacement and new technology additions

6.1.3 Biomedical Engineering—BE

Bronzino [2] defines BE is an interdisciplinary branch of engineering that ranges from theoretical, nonexperimental undertakings to state-of-the-art applications. In this regard the topic may involve research, development, implementation, and operation.

According to the International Federation of Medical and Biological Engineering—IFMBE (Harmonization of Biomedical Engineering Education—Status and Challenges, IFMBE, R. Magjarevic et al., 2016.) BE is the discipline that

A. advances knowledge in engineering, biology and medicine, and in basic sciences;
B. improves human health by design and problem solving skills of engineering;
C. is science applied to diagnosis, monitoring, therapy, and rehabilitation, but also to prevention and prediction;
D. integrates engineering sciences with biomedical sciences and clinical practice.

Some challenges that BE has are

- ICT in medicine and healthcare
- Minimally invasive surgery
- Biomedical sensors
- Medical imaging and visualization of the data
- Intelligent materials
- Cellular and stem cell engineering
- Nanotechnology
- Modeling and simulation of physiologic systems and human body as a whole

Biomedical Engineers are in charge of some of the following responsibilities (Kulhna University of Engineering & Technology—KUET, Department of Biomedical Engineering, MD Bashir Udin, Bangladesh, 2014):

A. Designing prostheses
B. Designing replacement parts for people
C. Creating systems to allow the handicapped to function, work, and communicate
D. Managing the technology in the hospital system

6.1.4 Clinical Engineering—CE

The American College of Clinical Engineering—ACCE (2016) states that CE education is based in classical engineering, supplemented with a combination of courses in physiology, human factors, systems analysis, medical terminology, measurement, and instrumentation.

ACCE remarks that CE professional is qualified by education to practice in the healthcare environment where technology is created, deployed, taught, regulated, managed, or maintained related to health services. An interesting aspect to observe is CE is an interdisciplinary field practiced in a variety of settings and presenting a diversity of challenges. In this regard for ACCE a clinical engineer, by education and training, is a problem solver, working with complex human and technological systems.

At the hospital, some of the functions of Clinical Engineers are as follows:

A. Participation in the planning process and in the assessment of health technology.
B. Assurance of regulatory compliance in the medical technology management area.
C. Investigation in incidents, and active participation in training and education of technical and medical personnel.
D. Contribution to the design, development, and management of medical, communications, and information systems.

Fig. 6.1 summarizes the definitions and scope of the topics presented.

The correct selection of Healthcare Workforce to be included in the programs of Education and Training affects the results expected. Attending to the rationality of the investment required (1) the definition of the needs; (2) the determination of the workforce to be involved in the programs; (3) the specific results expected from the programs; and (4) the way it is expected the programs of Education and Training will improve specific units of the healthcare organization, which are some important aspects to be considered in the short-, medium-, and long-term plans.

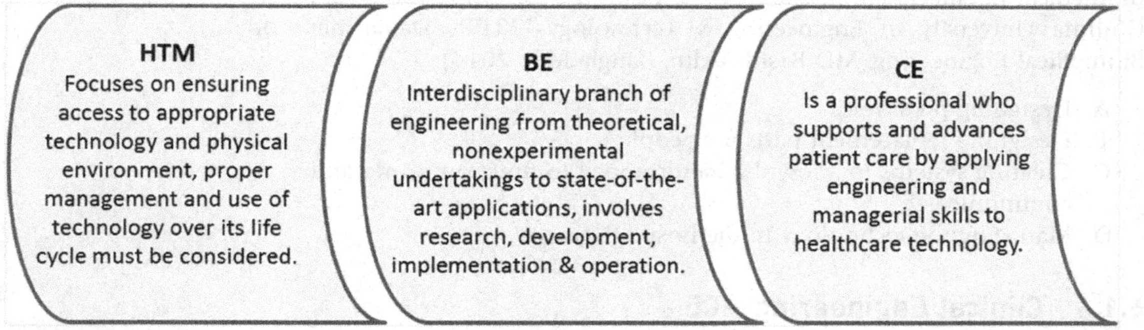

HTM
Focuses on ensuring access to appropriate technology and physical environment, proper management and use of technology over its life cycle must be considered.
BE
Interdisciplinary branch of engineering from theoretical, nonexperimental undertakings to state-of-the-art applications, involves research, development, implementation & operation.
CE
Is a professional who supports and advances patient care by applying engineering and managerial skills to healthcare technology.

FIGURE 6.1
Health Technology Management, Biomedical Engineering, Clinical Engineering definitions. *Elaborated by Rivas (2017).*

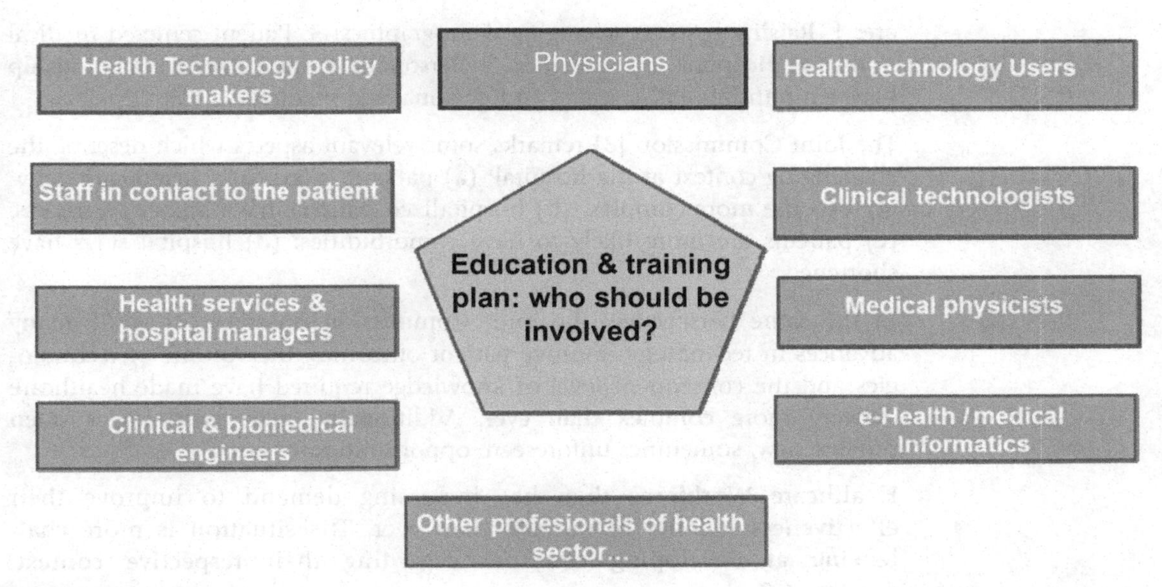

FIGURE 6.2
Health technology: education and training. *Elaborated by Rivas (2017).*

Some members of Health Workforce included in this regard are physicians, nurses, and healthcare staff in contact with the patient, health technology policy makers, health technology users, health service and hospital managers, health managers, clinical and biomedical engineers, medical physicists, clinical technologists, e-health/medical informatics, and others. See Fig. 6.2.

6.2 CHALLENGES AND TRENDS OF HEALTHCARE WORKFORCE

Some of the factors which characterized the context of health are described by Wheeler ("Thoughts of the Future", Bruce Wheeler, EMBS President, Department of Biomedical Engineering, University of Florida, 2015.): (a) Timeliness; (b) Revolutionary Medical Imaging; (c) Infrastructure of information processing; (d) Molecular biology, genomics...; (e) Spectacular diagnostics; (f) Critical need for technology for solutions at multiple levels.

In addition, CGI ("Healthcare Challenges and Trends: The patient at heart of care", Management & IT Consultant Group—CGI, 2014.) remarks globally, all health economies are facing similar challenges. The advent of new consumer technology is introducing even more challenges, or bringing older ones to the fore. Some of the challenges and trends for Healthcare CGI states are: 1. Raising costs; 2. Changing demographics; 3. Patient centered medical home; 4. Hospitals as networks; 5. Personalized medicine; 6. Relationship between translational research and personalized medicine.

The Joint Commission [3] remarks some relevant aspects which describe the challenging context at the hospital: (a) patients' needs and healthcare delivery become more complex; (b) hospitalized patients have higher awareness; (c) patients are more likely to have comorbidities; (d) hospital stays have shortened.

In the same perspective, the Joint Commission observes [4] while many advances in technology improve patient outcomes, the volume of technologies and the consequent level of knowledge required have made healthcare delivery more complex than ever. Additionally, new technologies often provide new, sometimes unforeseen, opportunities for error.

Healthcare Workforce then has increasing demand to improve their effectiveness, quality of outputs than ever. The situation is more challenging at developing countries regarding their respective context. See Fig. 6.3.

Thinking about the future of health organizations mostly the one related to the hospital, Ribera [5] emphasizes on the triple challenge context for the hospital: 1. Increase in healthcare needs; 2. Decrease in resources; and 3. Changing social values.

The author also proposes a framework for the hospital of the future based on five components:

1. Context: Political, Financial/Social Demographics, Technology, Environment, Legal, Market, Patients.
2. Leadership & Strategy: Vision, Mission, Values, Culture, Governance.
3. Resources & Capabilities: Personnel, Facilities, Financial, IT, Partnerships, and Others.
4. Processes: Business Processes: Purchasing, Reimbursement, Marketing and Commercialization, Knowledge Management, Information Management, Control & Finances, Budgeting, Innovation.
   Clinical Processes: Monitoring/Preventing, Diagnosing, Medication, Intervening/Treatment, Recovering/Rehabilitating, Aging, End of Life.
5. Results & Value Creation: Patients, Personnel, Payers, Society, Financial, Quality.

The framework facilitates to think and analyze the hospital as a system considering and understanding the diverse aspects which explain the complexity of the health organization. See Fig. 6.4.

Health technology at developing countries

- Limited medical device regulations.
- High percentage of devices that are out of service.
- Weak after sale device support with nearly all service from manufacturers or their representatives.
- A shortage of health technology staff in hospitals.
- Very limited maintenance budget.
- Little health technology support training.
- Limited technology management.

FIGURE 6.3

Health technology at developing countries. *Clark Tobey, et al. Health technology at developing countries. In: 2nd WHO—global forum on medical devices, Geneva, 2013.*

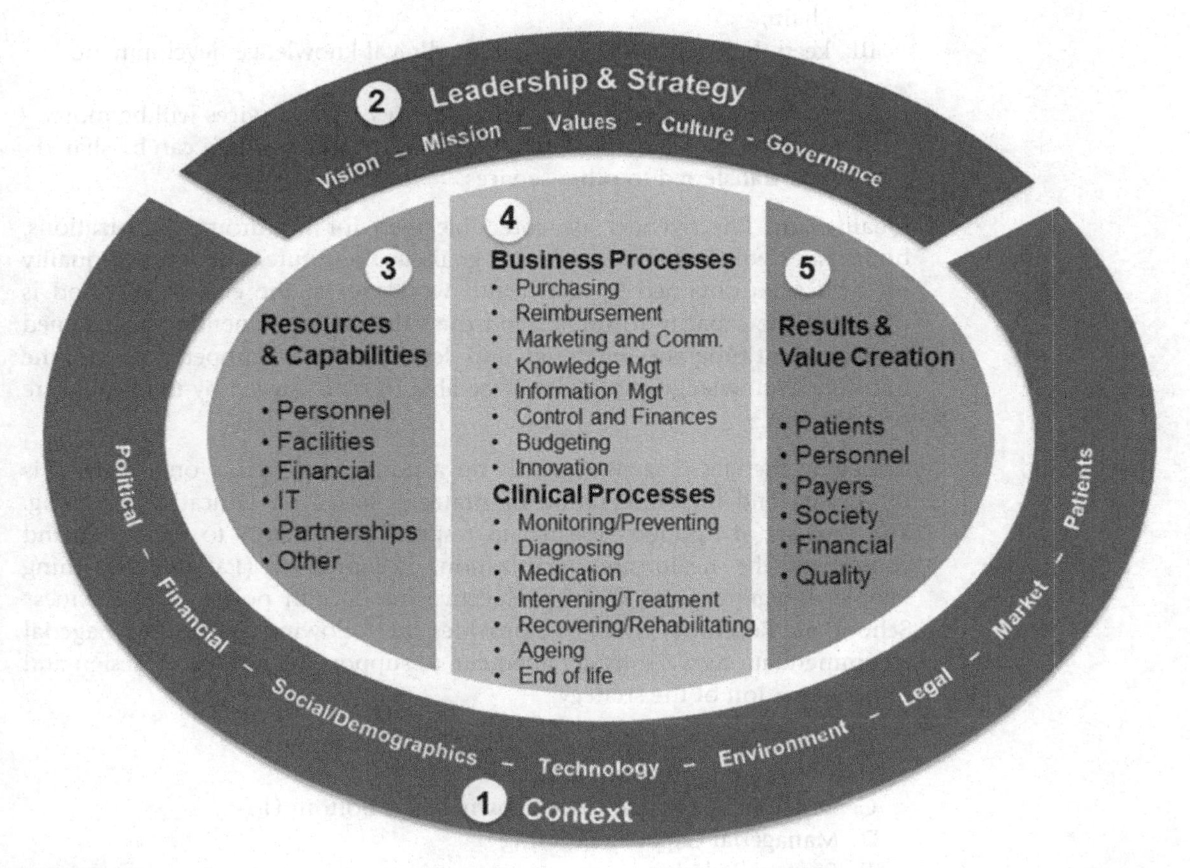

FIGURE 6.4

Hospital of the future framework. *Ribera J, et al. IESE, CHRIM, 2015 (See also: Malcolm Baldridge Criteria, https://www.nist.gov/baldrige/publications/baldrige-excellence-framework).*

6.3 HOSPITAL: FACING THE PRESENT AND BEING ALIGNED TO THE FUTURE

The following four insights provided by Ribera are oriented to support the hospital to be aligned to the future; in this regard he proposes hospitals should

I. have to find new ways to keep providing efficient and high-quality services, over the past few decades life expectancy has grown, public health interventions have more complexity, and the progress of medicine, as the public demand more and better healthcare services;
II. play an active role in helping public administration and society deal with a healthcare economic challenge bringing vision and knowledge to the debate on the configuration of a future healthcare system. They could establish or defend their key position in the healthcare value chain;
III. keep their role in aspects such as clinical knowledge development, overall health chain and system design;
IV. seek to protect synergic hospital services, these services will be more effective in generating knowledge and capabilities than can be shared and transferred to other services.

Quality and Effectiveness are core objectives for healthcare organizations, both issues will keep being relevant goals in the future. The level of quality of the results obtained by the health workforce at the end of a period is based on the capacity to understand the value of implementing a sustained improvement program, the design and development of competences, and the transfer of knowledge required and be able to be managed by the healthcare organization.

Achieving the stated goals depends on a number of factors one of them is the design and implementation of strategies based on Education, Training, and Projects of Applied Research to respond consistently to the needs and context of the healthcare organization. Rosenmoller (Patients becoming People: Integrating the social perspective into health policy, IESE Business School, M. Rosenmoller, 2016.) provides the following Practical Managerial recommendations we consider adequate to support the process of design and implementation of the strategy:

A. Stakeholder involvement and institutional support
B. Transparency/Information
C. Different approaches: Top-Down versus Bottom-Up
D. Managerial Capacity Building
E. Process Redesign
F. Leadership

6.4 IMPROVING QUALITY AND EFFECTIVENESS IN THE HEALTH ORGANIZATION: DEVELOPMENT OF SKILLS, TRANSFER OF KNOWLEDGE, AND EXCHANGE OF BEST PRACTICES

As discussed above Education and Training are key processes for the excellence and sustainability of healthcare organizations. Education and Training can also play a significant role (HealthCare Education and Training, Rand Corporation, 2017) in determining healthcare professionals' proficiency with the most recent interventions and technologies.

Related to Safety some reasons to invest on Education and Training are (Health Technology Education and Training for Health Workforce, T. Clark, 2010): (i) User error is a common cause of adverse events; (ii) Knowledge equals Safety; (iii) Safety education is not only about hazards; (iv) Joint Commission; (v) Education can reduce health technology adverse events and improve Safety.

6.4.1 Training

Two types of Training: On the Job and Off the Job are available for Healthcare Workforce [6]:

6.4.1.1 Training On the Job

The knowledge and practice is demonstrated (often in stages) and then undertaken by the trainee under supervision. Since the employee is trained while he is doing his functions, his knowledge will be improved. The process may be complemented by training manuals which are expertly written documents. On the job training involves the employee who may be disturbed by daily work or colleagues and this type of training is commonly used in United States.

6.4.1.2 Training Off the Job

The training is applied on a location different from the place of work of the staff. After the course the employee must implement the contents that were learnt on his functions. In this type of training the employee can focus 100% on the course.

6.4.2 Strategies and Tools for Assessment of Needs

Bauld (User Training, Thomas Bauld, Advancing Safety in HealthCare Technology, 2009) suggests the following Training Strategies:

1. Train the Trainer
2. Direct User Training

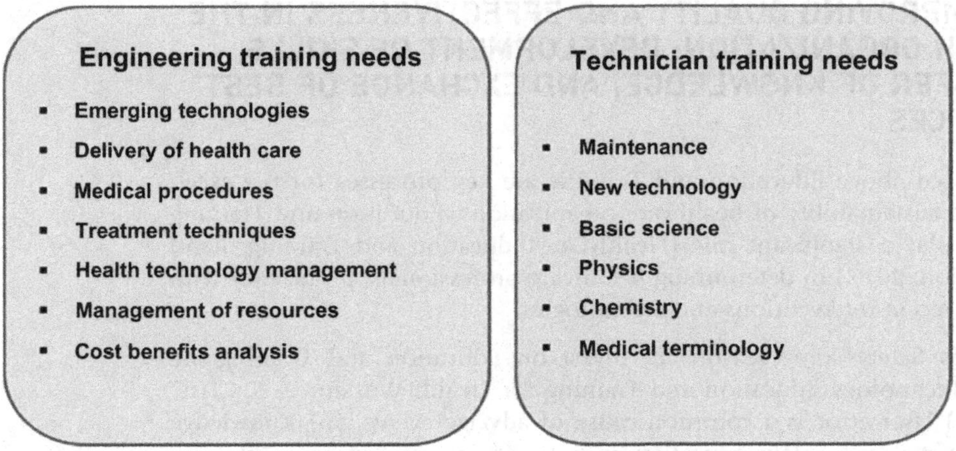

FIGURE 6.5
Education, training & competency assessment. *Clark T. Healthcare Technology Foundation, 2011.*

3. Self-instruction—media from manufacturers or professional associations
4. Videoconferencing
5. Internet and computer based

Some tools to assess the needs of Training at the Healthcare organization are provided also by Bauld: Surveys, Interviews, Incident Reports, Comments from Unit Rounds, Problem Descriptions form Work Orders. Fig. 6.5 shows some examples of topics for Training related to engineers and technicians at the healthcare organization.

6.4.3 Building Capacities of Health Workforce: A Strategic and System Approach

"Linkage and Leapfrog Strategy" a sustained strategy-mix designed by CENGETS (Health Technopole CENGETS; http://its.uvm.edu/PUCP_CENGETS/HomePage.html) and applied to Educational and Training Programs. The strategy is supported by a Projects & Problems Based Learning Methodology [7].

Some results obtained are (i) Development of Best Practices through Applied Research Projects at Peruvian health sector and (ii) Promotion of the profession of BE in Peru. Key components of the model are Innovation, Technology, Management, Education, Interaction, Leadership, Multidisciplinary Perspective, and International exchanges. See Fig. 6.6.

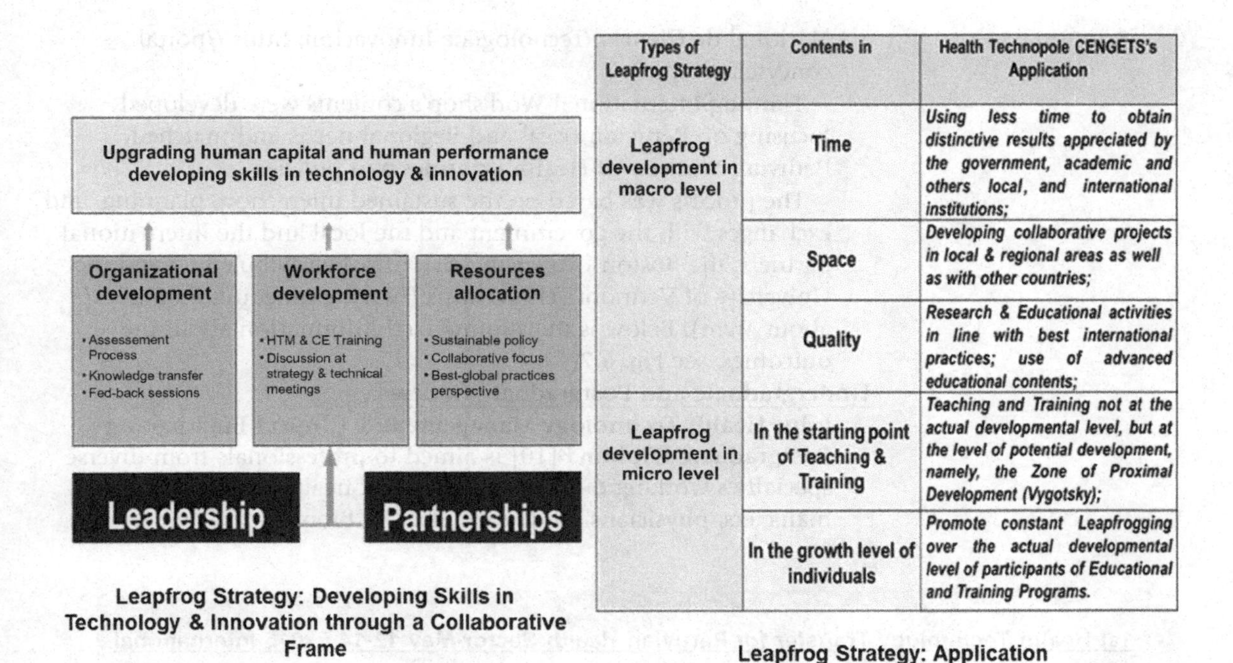

Types of Leapfrog Strategy	Contents in	Health Technopole CENGETS's Application
Leapfrog development in macro level	Time	*Using less time to obtain distinctive results appreciated by the government, academic and others local, and international institutions;*
	Space	*Developing collaborative projects in local & regional areas as well as with other countries;*
	Quality	*Research & Educational activities in line with best international practices; use of advanced educational contents;*
Leapfrog development in micro level	In the starting point of Teaching & Training	*Teaching and Training not at the actual developmental level, but at the level of potential development, namely, the Zone of Proximal Development (Vygotsky);*
	In the growth level of individuals	*Promote constant Leapfrogging over the actual developmental level of participants of Educational and Training Programs.*

Leapfrog Strategy: Application

FIGURE 6.6
Health Technology Management & Clinical Engineering: a strategic model in Peru: Building & Strengthening capacities for HTM & CE at developing countries. *Vilcahuamán L, Rivas R. WHO—global forum to improve developing country access to medical devices, Bangkok, 2011.*

Below are some selected activities developed in Peruvian Health sector, and they are related to the implementation of Linkage and Leapfrog Strategy; a key factor was to maintain consistency with the Peruvian context, their needs and to apply a multidisciplinary and global approach [8]:

A. **Educational activities including best international practices; Use of advanced educational contents.**

International Workshops:

- **"Health Technology Transfer for Peruvian National Institute of Health"; "Health Technology Transfer for Epidemiology Research and Public Health in Heavy Metals" International Workshops** [9] were demanded by National Institute of Health—NIH (Instituto Nacional de Salud—INS, Ministerio de Salud de Perú; http://www.portal.ins.gob.pe/es/) from the Ministry of Health of Peru. The process was sponsored and funded by the National Council of Science, Technology and Innovation of Peru—CONCYTEC (Concejo

Nacional de Ciencia, Tecnología e Innovación; https://portal.concytec.gob.pe/).

Training International Workshop's contents were developed focusing on Peruvian Local and Regional needs and matched Peruvian Ministry of Health priorities and policies.

The process was based on the sustained interaction, planning, and exchanges with the government and the local and the international partners: the Boston University (BU; http://www.bu.edu/) and the University of Vermont (UVM; https://www.uvm.edu/about_uvm/about_uvm). Below is the summarized information about the outcomes, see Fig. 6.7.

Undergraduate and Postgraduate Courses:

- **Joint Health Technology Management & Clinical Engineering Postgraduate Diploma** [10] is aimed to professionals from diverse specialties working on private and public health sector: engineers, managers, physicians, biologists, etc. The Biomedical Engineering

1st Health Technology Transfer for Peruvian Health Sector-May 12-14 / 2015 International Workshop:

- ***Formulation of clear and comprehensive policies in: health innovation, HTM & HTP and Health Technology Transfer in Peruvian health sector:*** *Six HT Policy Guidelines were improved with the participation of the board: MoH, NIH, CENGETS and the Univ. of Vermont: Professors Barry Finette, Tobey Clark, Asim Zia.*
- ***Improvement of capacities for the Implementation of Health Technology Transfer Offices in Peruvian health sector:*** *Health staff in charge of research, management and policy development of the institutions and agencies of MoH, Departments of Health and Regional governments, Universities, Research institutes, Technological centers and companies and other institutions engaged in research and development and health technological innovation were trained.*

1st Health Technology Transfer for Epidemiology Research and Public Health in Heavy Metals-May 6-8 / 2015. International Workshop:

- ***Publication on NIH Scientific Review of an Article with Topics and Findings of the International Workshop.*** *Led by NIH, CENGETS and Prof. Herbert Voigt to be promoted in Latin and Center America.*
- ***Applied Research of Development of Devices for Infection Detection:*** *Led by NIH, BE Master Program, CENGETS & Prof. Herbert Voigt, with the contribution of Dr. Cesar Cabezas, former NIH's Director.*

FIGURE 6.7
Health technology transfer for Peruvian NIH—Workshop outcomes. *Rivas R, et al. WHO-PAHO-IFMBE-ACCE—Health technology management seminar, Denver, USA, 2015.*

Master Program of Pontificia Universidad Catolica del Peru—PUCP (www.pucp.edu.pe), Medicine School of Universidad Peruana Cayetano Heredia—UPCH (www.upch.edu.pe) and CENGETS lead the program which finishes with projects applying the contents of the course.

- **Joint Biomedical Engineering Undergraduate Program** [11] between PUCP University and UPCH University started processes for the creation of the program on 2011, the academic project was finished on 2015; at the end of 2016 both universities announced and promote the new career. The studies will start on 2017. PUCP Biomedical Engineering Master Program, CENGETS, UPCH School of Sciences, UPCH School of Medicine, the University of Boston, and the University of Vermont were some of the collaborators on the process of creation of the academic program [12].
- **First Health Technology Management & Health Technology Planning and Health Technology Transfer Course** [13] responded a request of National Institute of Health—NIH of Peru; 45 professionals of different units such as National Laboratories, Public Health Center, and Technology Transfer & Research Office participated in the course on November–December 2015. PUCP Biomedical Engineering Master Program and CENGETS led the program, international collaborators were the Panamerican Health Organization—PAHO (http://www.paho.org/hq/index.php?option=com_content&view=article&id=91%3Aabout-paho&lang=en), American College of Clinical Engineers—ACCE (http://accenet.org/Pages/Default.aspx), International Federation for Medical and Biological Engineering—IFMBE (http://2016.ifmbe.org/about-ifmbe/.), the University of Vermont and the HealthCare Technology Foundation (Healthcare Technology Foundation—HTF: Advancing Healthcare through the use of Safe and Effective Technologies; http://www.thehtf.org/default.asp).
- **Health Technology Innovation: Health Technology Management & Planning Online Course** [14] aimed to Chemical & Pharmaceutical Peruvian College—CQFP (Colegio Químico Farmacéutico del Perú—CQFP; http://www.cqfp.org.pe/), the contents were developed by Prof. Clark and Prof. Rivas, they were aligned to Health Technology and Health Innovation global trends, Peruvian Ministry of Health policies and strategic objectives of health sector on private and public market; 25 professionals from Ministry of Health, Peruvian enterprises and Nonprofit organizations participated. The course was the First Online course implemented for CQFP. The program was co-led by Prof. Clark of the University of Vermont

Joint Health Technology Management & Clinical Engineering Postgraduate Diploma:
Pontifical Catholic University of Peru-PUCP Biomedical Engineering Master Program, UPCH Medicine School and CENGETS lead currently the 3rd HTM & Clinical Engineering PUCP-UPCH 2017 Joint Diploma, the 1st one finished with an international training in Mexico at the Ministry of Health-Excellence Technology National Center-CENETEC, the 2nd finished with an international training in Colombia at the Ministry of Health.

Joint Biomedical Engineering Undergraduate 2017 Program:
The 2 top Peruvian leaders in the Academia: Pontifical Catholic University of Peru-PUCP-Engineering and Universidad Peruana Cayetano Heredia-UPCH-Medicine created a Biomedical Engineering Undergraduate Program. The Biomedical Engineering Master Program and CENGETS participated on the commission, Professors Herbert Voigt from the Univ. of Boston and Tobey Clark from the Univ. of Vermont are sustained collaborators. The program will start at the end of March 2017.

1st Health Technology Management & Health Technology Planning and Health Technology Transfer 2015 Course for NIH:
13 Proposals developed by work teams for applied research based on the topics of the course: NIH leads the continuation of the activities led by PUCP BE Master Program and CENGETS. International collaborators of the course were: Panamerican Health Organization-PAHO, ACCE & IFMBE, the Univ. of Vermont and Healthcare Foundation.

Health Technology Innovation: Health Technology Management & Planning On-Line 2016 Course for Chemical & Pharmaceutical Peruvian College-CQFP:
4 Proposals developed by work teams for applied research based on the topics of the course. Professionals from Ministry of Health, Non profit organizations, Private enterprises participated on the course. Contents designed by Professor Clark from the University of Vermont and Professor Rivas from CENGETS

FIGURE 6.8
Peruvian health sector educational outcomes [18].

and PAHO Online courses author and consultant [15] and Prof. Rivas of CENGETS who collaborated with Prof. Clark developing contents and teaching on some virtual courses as consultant for PAHO.

Fig. 6.8 summarizes information related to the outcomes of these courses:

B. **Developing collaborative projects in local & regional areas in collaboration with other countries: Applied Research projects.**
   - **Healthcare Technology Management—HTM Comprehensive System in INMP [16]**

In the context of new technologies, it is not enough to improve the outdated model, but a new model needs to be raised by re-engineering to achieve appropriate levels of clinical effectiveness, efficiency, safety, cost control, and quality that users expect on the technology used in hospitals. Process map developed in the Comprehensive System of healthcare technology management has proven to be suitable to incorporate gradually and without major disruption to the current processes of the hospital, both aspects determine a feasible application, whether partial or complete. The Process Map is a guide to effectively update HT and improves the functionality of clinical services. One of the findings shows that National Institute of Maternal and Perinatal Health—INMP (Instituto Nacional Materno Perinatal—INMP; http://www.inmp.gob.pe/) tend to re-think their organization according to the proposed processes; this fact supports the proposal of an organizational structure of higher quality than before. The indicators feedback on-time the state of technology in the hospital, they also facilitate the effective and on-time correspondent action. The computer platform developed in HTM Comprehensive System proves to be flexible and scalable, allowing high feasibility of application due to the easy adaptability to particular environments in hospitals.

- **Implementation of Health Technology & Clinical Engineering Unit in Peruvian Hospitals** [17]

  Hospital adequate framework implies the management of technology with a high-quality level of performance and integration. The process needs to be transverse to the "traditional-dynamic" of Peruvian hospital organization but requires a strong collaboration between the medical branch and the administrative branch. National Institute of Maternal and Perinatal Health—INMP and CENGETS established a strategic alliance to collaborate, share, and merge capacities to support the implementation of a Management Unit in Health Technology and Clinical Engineering in INMP. CENGETS brings the ongoing monitoring and validation of methods, procedures, processes, and protocols to the unit's performance. CENGETS intervention involves the management of collaboration from international experts and partners: PAHO/WHO (World Health Organization—WHO; http://www.who.int/en/.), HTAi (Health Technology Assessment International—HTAi; http://www.htai.org/htai/about-htai.html), University of Vermont and ORBIS International (ORBIS International; http://www.orbis.org/). INMP and CENGETS inter-institutional collaboration promotes and sustains the integration of capacities focused on Organization, Processes, Education, and Leadership. Some outcomes are (a) Classification of biomedical equipment focused on

Patient Safety; (b) Evaluation, assessment, and reporting on the impact of technology for the acquisition of 1 and 1 pulse oximeter and 1 neonatal ventilator; (c) Creation and design of software Biomedical Equipment Maintenance; (d) Strengthening INMP TeleHealth System in the areas of Clinical Engineering and Health Technology Management Networking with the regions Trujillo and Chimbote.

References

[1] Cook D, Sandham J. Improving healthcare technology to deliver transformation, 2016.

[2] Bronzino J. Medical devices and systems. The biomedical engineering handbook. 3rd ed. 2006.

[3] Joint Commission. Guiding principles for the development of the hospital of the future. Quality in Health Care; Illinois, US, 2008.

[4] Joint Commission. Joint commission on accreditation of healthcare organizations, 2012.

[5] Ribera J, et al. Hospital of the future. IESE—Center for Research in Healthcare Innovation Management—CRHIM, Barcelona, Spain, 2016.

[6] FAKT. Healthcare technology: training skills for hospital, consult for management, training & technologies, 1999.

[7] Vilcahuamán L, Rivas R. A strategic model in Peru: building and strengthening capacities for Healthcare Technology Management & Clinical Engineering aimed at developing countries. 1st Global forum on medical devices. Bangkok, Thailand: World Health Organization—WHO; 2011.

[8] Vilcahuamán L, Rivas R. Health technopole: innovation applied to clinical engineering & health technology management education. In: 32nd IEEE annual international conference, engineering in medicine and biology society—EMBS: "merging medical humanism & technology", doi:10.1109/IEMBS.2010.5626461, 2010.

[9] National Institute of Health of Peru, Boletín Nacional, Año 21, Nro. 5 y 6, May–June, ISSN: 1606-6979, 2015.

[10] PUCP-UPCH. Joint health technology management & clinical engineering postgraduate diploma. <http://files.pucp.edu.pe/posgrado/wp-content/uploads/2016/10/31150050/Gesti%C3%B3n-de-tecnolog%C3%ADa-en-salud-e-ingenier%C3%ADa-cl%C3%ADnica-1.compressed.pdf>.

[11] PUCP-UPCH. Joint biomedical engineering undergraduate program.

[12] Rivas R, et al. Biomedical engineering education in peru in 2015: a unique and innovative collaboration in latin america. In: World Congress on Medical Physics & Biomedical Engineering, Toronto, Canadá. International Federation of Medical and Biological Engineering—IFMBE Proceedings. ISBN: 978-1-988006-00-0; 2015.

[13] Rivas R. 1st Health technology management & health technology planning and health technology transfer course, Academic coordinator final report, 2015.

[14] Rivas R, Clark T. Health technology innovation: health technology management & planning on-line chemical & pharmaceutical peruvian college—CQFP course, Professors final report, 2016.

[15] Clark T, Rivas R, et al. Biomedical technology online courses for the Americas. In: World congress on medical physics & biomedical engineering, Toronto, Canadá. International Federation of Medical and Biological Engineering—IFMBE proceedings. ISBN: 978-1-988006-00-0; 2015.

[16] Vilcahuamán L, Córdova M, Kafalatovich J, Rivas R. A Comprehensive System for Healthcare Technology Management—HTM, 2016. To appear in the International federation of medical and biological engineering—IFMBE proceedings. Latin American conference in biomedical engineering—CLAIB. ISSN: 1680-0737.

[17] Vilcahuamán L, Rivas R, Portella J, et al. Health technology management & clinical engineering unit in peruvian hospitals: technology excellence and quality at INMP, International Federation of Medical and Biological Engineering—IFMBE proceedings, Latin American conference in biomedical engineering—CLAIB. ISBN 978-3-642-21198-0; 2011. p. 579−82.

[18] Rivas R, Vilcahuamán L. Global Clinical Engineering Day: CENGETS Peru, World Health Organization—WHO, American College of Clinical Engineering—ACCE, International Federation for Medical and Biological Engineering—IFMBE, Health Technology Foundation—HTF, October 2016.

CHAPTER 7

Applied Research & Innovation in Healthcare Technology

Applied research, an underutilized capacity to renew and innovate technology in hospitals.

"Research can be functionally divided into basic (or pure) research and applied research. Basic research is usually considered to involve a search for knowledge without a defined goal of utility or specific purpose. Applied research is problem-oriented, and is directed towards the solution of an existing problem. There is continuing controversy over the relative benefits and merits to society of basic and applied research. Some claim that science, which depends greatly on society for its support, should address itself directly to the solution of the relevant problems of man, while others argue that scientific inquiry is most productive when freely undertaken, and that the greatest advances in science have resulted from pure research. It is generally recognized that there needs to be a healthy balance between the two types of research, with the more affluent and technologically advanced societies able to support a greater proportion of basic research than those with fewer resources to spare. Yet another way of classifying health research, be it empirical or theoretical, basic or applied, is to describe it under three operational interlinked categories of biomedical, health services and behavioral research, the so-called health research triangle. Biomedical research deals primarily with basic research involving processes at the cellular level; health research deals with issues in the environment surrounding man, which promote changes at the cellular level; and behavioral research deals with the interaction of man and the environment in a manner reflecting the beliefs, attitudes and practices of the individual in society" [1].

It is true that some hospitals are investigating, usually applied research and few others are involved in basic research, but it is also true that the participation of biomedical engineers or clinical engineers is very small. We want to outline this scenario, sure to have multiple factors, but what is being gained? and what is being lost? Why should a hospital do research? We must start from the beginning. The role of the hospital is to provide individual and

Healthcare Technology Management Systems. DOI: http://dx.doi.org/10.1016/B978-0-12-811431-5.00007-2
© 2017 Elsevier Inc. All rights reserved.

collective health services to the community, with quality of service, clinical effectiveness, safety, efficiency, and cost control. We have also proposed that any successful health initiative requires at least three components, a medical branch, an administrative branch, and a branch of technology. Here we must remember (see Chapter 1: Healthcare Technology Management (HTM) & Healthcare Technology Assessment (HTA)) that health technologies are clinical technologies (clinical procedures, drugs, and medical devices), support technologies (infrastructure, hospital equipment, information systems, etc.), community health technologies, (prevention, protection, and promotion), and environmental health technology. In this scenario, it should also be considered that large hospitals often have the role and capacity to conduct research, however, small hospitals should not rule out conducting research or participating in research in conjunction with higher resolution hospitals because it is about improving existing processes or validating technologies. The following are the reasons why research, at least applied research or health research, would be necessary and timely in a hospital:

1. The hospital must adapt to the high dynamism in the generation of health technology, to take advantage of the benefits of its improvements in effectiveness, efficiency, and safety, for the benefit of patients, but part of this process needs to be investigated to know how to properly adapt to these new technologies.
2. Applied research in health also involves aspects of the medical branch, administrative branch, and the technological branch, so it is naturally multidisciplinary where the participation of biomedical engineers, clinical engineers, and similar others is relevant.
3. Without research, hospitals tend to be lagging and obsolete, unable to manage their own technology and unable to adapt to new ones. The role of biomedical engineers and clinical engineers is crucial for the technological development of hospitals.
4. The equipment is installed; however, the technology requires a complete process of incorporation. Knowing that technology is not only medical equipment, but a series of components such as related clinical procedures, infrastructure, information systems, and support of hospital equipment, it is necessary for the leadership of the engineers in hospitals and their capacity of design, management, and research to ensure an appropriate process of technology incorporation.
5. The continuous improvement of technology goes through an applied research process. Not everything is in the norms and regulations, the publications may not be enough, besides not everything is known. Thus, in many cases the improvements can be done from applied research or validated experimentation.

6. Better designs improve the quality of technology for health. The ability to design promotes the ability to research and generate improvements in the current state of technology in hospitals.
7. A hospital with the capacity to improve and create technology has great potential to provide better quality of health services. Better technologies (see Chapter 1, Healthcare Technology Management (HTM) & Healthcare Technology Assessment (HTA): procedures, medical devices, materials, assistive technologies, and community health technologies) arise from applied research.
8. By having applied research capacity, it is likely to enter into the development of technological innovations, which entails having the capacity for basic research.
9. There is a hidden capacity in hospitals to be generators of technology useful for their community. We think it is not contradictory to the role of the hospital, which we normally see as providing health services, but this can be strengthened by new technologies that the hospital itself can generate or can adapt from foreign technologies.
10. Research in hospitals offers the possibility to generate technology and innovate, as well as to improve standards and regulations, and therefore to be a promoter of a viable biomedical industry.
11. A researching hospital tends to attract better professionals, singularity biomedical engineering and clinical engineering professionals, because of its better professional development perspective. Engineering professionals typically find little space for professional development in hospitals, which discourages them, however, the research offers much better prospects.
12. The cost of doing research can be covered with the ability to generate funded projects, patents, and publications, as well as transfer technology for successful results.
13. The more efficient the internal processes of the hospital as a result of research, the possibility of having resources to research is obtained, but in turn the research offers the possibility of improving the internal processes, therefore, it is a virtuous circle that is due encourage.
14. Basic and applied research in hospitals can be done in conjunction with research institutes, universities, companies, and government. This is an activity in which biomedical engineering and clinical engineering professionals can contribute and lead for the benefit of community health.
15. Research capacity in a hospital strengthens the capacity for professional training and training of new researchers. This is a capability of highly prized by universities and industry in general, but ultimately reciprocates for the benefit of patients.

Current research in hospitals should be open to all types of health technology (see Chapter 1: Healthcare Technology Management (HTM) & Healthcare Technology Assessment (HTA), Fig. 1). Although we find epidemiological research, research in health policy, prevalent diseases, drug use, health according to age group, behavioral health, etc., from the engineering point of view, we still have many issues to research. In the first place, it can be said that all the mentioned researches in turn have a technological environment that must equally be researched and generated. Instrumentation is required to measure the parameters and it is necessary to create environmental conditions for the management of biological components; biocompatible materials, automatic machines, mathematical models, software, process design, etc., are required. It is very good to promote multidisciplinary research, but not only to consider specialists in the field of medicine and biology, must also include biomedical engineers, clinical engineers, medical physicists, hospital architects, among others, which in turn will be done in double specialists, that is to say they will dominate some field of their profession but also they will become specialists in the fields of medicine and biology required for research initiatives. In this scenario we observed that the hospital organization, which planted the formal margins of action of professionals, should be reviewed. Typically, the role of researching from hospital engineering departments is not planned, because they are defined for the purpose of medical equipment maintenance, leaving aside many other functions that a biomedical engineering and clinical engineering professional should perform. Once again, we observe that the main constraints to these new functions, such as applied research, are the health policies and the organization of hospitals established for contexts of more than 6 or 7 decades.

In the next subchapters we will outline some criteria and premises that can serve to incorporate this important research work, at least applied research, into the routine practice of biomedical engineers and clinical engineers in hospitals.

7.1 IDENTIFYING HOSPITALS TECHNOLOGY ISSUES TO RESEARCH

Identifying technology needs at the hospital, seen in Chapter 4, Health Technology Planning and Acquisition, is a good starting point for identifying applied research options. It is true that the needs and problems found in the vast majority can be solved with the expertise of professionals, however, there will be some that fall within the scope of research, some in the field of applied or experimental research and few others can even give light on alternatives for basic research. What does distinguish a requirement that can be solved with the expertise of engineering professionals, from another requirement that requires a research process? In the first place, we must have an institutionalized research policy that promotes it, indicating that the research will be appreciated and valued, and that the established institutional function must be in place, which in the long run will create the minimum conditions to research, specialty by engineering staff. Since the institutional plans of hospitals set goals for health problems to solve, it will now be necessary to decompose them to identify the related technological aspect that may require a research, the impact of which will help to solve the health requirement.

Health research follows clear scientific foundations [1]:

1. Order: The organized observations of entities and events can be classified according to their common properties and behaviors.
2. Inference and possibility: The statements and conclusions should be accepted because the evidence makes one or more premises true. Inferences can be deductive or inductive. In the inductive inference one has the possibility that the premises may be true but the false conclusions.
3. Probability assessment: The critical requirement in the design of research, the one that ensures validity, is the evaluation of probability from beginning to end. The most salient elements of design, which are meant to ensure the integrity of probability and the prevention of bias, are representative sampling, randomization in the selection of study groups, maintenance of comparison groups as controls, blinding of experiments and subjects, and the Use of probability (statistical) methods in the analysis and interpretation of outcome.
4. Hypotheses: Hypotheses are carefully constructed statements about a phenomenon in the population. The hypotheses may have been generated by deductive reasoning, or based on inductive reasoning from prior observations. One of the most useful tools of health research is the generation of hypotheses which, when tested, will lead to the identification of the most likely causes of disease or changes in the observed condition. In health research, hypotheses are often constructed and tested to identify causes of disease and to explain the distribution of disease in populations.

Naturally the hypothesis must be tested or validated, for which it is necessary to design the study. If an epidemiological approach is followed, it will be based on statistical principles, for which we have the case-control studies, cohort studies and retrospective cohort studies, and the experimental type study. The research should then be carefully planned and executed considering the integrity of one's self and the value of the scientific method, in addition to maintaining critical thinking and objectivity. Finally, the conclusion is elaborated based on the obtained evidence, which will qualify the validity of the hypothesis, that is, to what extent the supposed premises have been corroborated with the results achieved.

On the other hand, we have the focus of engineering and technology based on applied science. We must distinguish them from other approaches in the training of engineers, some more practical that emphasize the development of skills such as communication, creativity, critical thinking, business, and project manager, i.e., they work at a macro level systems. Both approaches are valid and universities try to take an appropriate balance according to their objectives and context. However, for the purposes of this chapter, we are more interested in the first approach, the engineer trained in a deep understanding of applied sciences, mathematics, physics, chemistry, i.e., engineering sciences and physiology as is the case for engineers in hospitals, capable of covering specific details. Then comes the concept of the engineer who designs, who can contribute in the generation of technology. It is said that in basic research, it does not need any design elements at all; project research typically contains at least some design; and consulting may be entirely design. Most importantly, the applied scientist is by definition not interested in changing reality, only in studying, explaining, and predicting it. The engineer, on the other hand, accepts the explanations and predictions of the applied scientist, and then acts to change reality to achieve some goal. Coarsely put, engineers design but applied scientists do not [2].

We believe that for the appropriate development of technology in hospitals, it is required to have applied research capacity for the study of reality, for example a deep study of healthcare technology assessment (HTA), cost effectiveness analysis for a health intervention, and a meta-analysis have already characteristics of an applied research, but also it is necessary to change the current conditions, that is to say, we need to act, and this leads us to have the capacity to design, which is characteristic of engineering. We require biomedical engineers and clinical engineers with the ability to design clinical procedures in conjunction with physicians, capable of systematizing metrological trials, designing test protocols, designing technological assistance devices for disability and rehabilitation, designing disinfection and sterilization procedures for reuse of accepted disposable devices, etc., but there is also an inescapable task which is to improve what already exists, seek greater clinical effectiveness, greater efficiency, better safety and if it is possible, to reduce costs.

The identification of cases to be studied or problems to be solved needs to be prioritized based on institutional objectives and plans, the feasibility of realization in terms of human, material, and financial resources, and the interest of the researcher or designer. The process of identifying cases can be approached from a general approach of hospital outcomes, for example the evaluation of adverse events in patients, or from a particular approach, such as within a clinical service or specifically a type of clinical procedure, for example to improve body temperature control for neonatal surgery. We believe that due to the intrinsic complexity of every hospital, it will always offer a wide range of possibilities to research and develop, but we are also in a world of budgetary constraints, so this work must be well planned. Thus, the research and development plan (included designs) must have clear and feasible objectives to achieve in a specific time. It should include the following components [1]:

1. Defining the intended role and scope of the research undertaking.
2. Determining the capabilities and resources of the research unit, to include personnel, facilities, equipment, supplies, time and budget, and accessibility of research material.
3. Selecting the research topic, considering factors such as
   a. magnitude of the problem and its impact;
   b. urgency of the need for a solution;
   c. relevance to the aims of the funding agency;
   d. amenability of the problem to investigation;
   e. feasibility of the approach;
   f. chances of success;
   g. expected impact of a successful outcome;
   h. spin-off in terms of training of staff and other research capability strengthening elements.
4. Constructing research protocols which will serve the guiding documents for the execution, monitoring, and evaluation of the research.
5. Setting up a well-defined administrative structure with lines of direction, supervision, consultation, and collaboration based upon task-specific job descriptions.
6. Formulating a schedule of targets for consolidation of results and preparation of these results for dissemination, including publication in the scientific literature.

On the other hand, the research proposal follows the following basic components, which will allow to formulate the problem, to plan the research, to execute the activities within the established objectives, and will lead to finding the solution to the problem in case of applied research:

1. Conceptualizing the problem:
   a. Identifying the problem (what is the problem?)
   b. Prioritizing the problem (why is this an important problem?)

   c. Rationale (can the problem be solved, and what are the benefits to society if the problem is solved?)
2. Background:
   a. Literature review (what do we already know?)
3. Formulating the objectives:
   a. Framing the questions according to general and specific objectives
   b. Developing a testable hypothesis to achieve the objectives
4. Research methodology:
   a. Defining the population, characteristics of interest and probability distributions
   b. Type of study (observational or analytical, surveys or experiments)
   c. Method of data collection, management and analysis:
      i. Sample selection
      ii. Measuring instruments (reliability and validity of instruments)
      iii. Training of interviewers
      iv. Quality control of measurements
      v. Computerization, checking and validating measurements
      vi. The issue of missing observations
      vii. The Statistical summarization of information
      viii. The Testing of hypothesis
      ix. Ethical considerations
5. Work plan:
   a. Personnel
   b. Timetable (who will do what, and when)
   c. Project administration
6. Plans for dissemination:
   a. Presentation to authorities to implement the results of the research (if applicable)
   b. Publication in scientific journals and other works (including those of the agency that funded the project) for wide distribution of the research findings
   c. Propose a technology transfer or marketing plan
7. Executive summary, giving an overview of the above topics in clear and simple language understandable by lay persons, and a list of references.

7.2 TECHNOLOGICAL INNOVATION FOR QUALITY IMPROVE IN HOSPITALS

"Innovation can be defined as—the intentional introduction and application within a role, group, or organization, of ideas, processes, products or procedures, new to the relevant unit of adoption, designed to significantly benefit the individual, the group, or wider society [3]. This definition is largely accepted among researchers in the field, as it captures the three most important characteristics of innovation: (a) novelty, (b) an application component, and (c) an intended benefit. In line with this definition, innovation in healthcare organizations are typically new services, new ways of working and/or new technologies. From the patient's point of view, the intended benefits are either improved health or reduced suffering due to illness" [4]. In 2005, industryweek [5] did a study about the effects of innovation on a company and they found that—overall revenue growth (78%), customer satisfaction (76%), growth in revenue from new products or services (74%), increased productivity (71%), and earnings/profit margins (68%) were a result of the impact of innovation efforts.

If we consider different types of technology in hospitals, clinical, support, prevention, protection, promotion, and environmental technologies, we find a wide range of possibilities for design, applied research, and innovation to be made by professional engineers. Improvements over what already exists can be considered as nondisruptive, incremental, or sustainable innovations, i.e., innovations that create better or new resources and processes. Disruptive or radical innovations are those that break old systems, create new markets, and call new actors, and in this case offer patients, an added value to which they must adapt. Telemedicine is an example of that, if it is raised about the old structures and processes most likely to have failures, however, implemented on a new vision of service, with new processes is likely the success and desired impact on the population. Organizational innovation is the introduction and intentional application of new policies, processes, products, and procedures for relevant clinical services, designed for the significant benefit of patients, for which aspects of the medical branch and the administrative branch should be considered as well as the branch of technology. Process innovation implies a significant improvement in healthcare services, with new or improved procedures, new technologies (architecture, equipment, software, etc.), and particularly information technologies have revolutionized health due to its capacity of transmitting data and images in a global way. Information technologies at least have four major impacts on health: they allow more services by third parties, for example tele radiology; The integration of information systems, both inside the hospital and externally, for example electronic and shared medical histories between different regions and even between countries; Monitoring the use of drugs on a global scale, aimed at sharing information between different health systems and countries; And the highest quality of information for physicians and patients through web pages, online publications, and specialized applications.

Multiple innovations in health are already underway, for example the use of robotic devices in rehabilitation therapies for victims of cardiovascular accidents, radio-stereotactic surgery without incisions for tumors inoperable according to traditional methods, and ablation therapy that can stabilize an irregular heartbeat. In all this we must consider the stakeholders, each one has different expectations regarding innovation: naturally, health professionals expect to improve clinical outcomes, improve diagnosis and treatment; Patients hope to improve their experience as patients, improve their well-being, reduce recovery time, and avoid delays or waits; Organizations expect to improve the efficiency of internal operations, reduce costs, increase productivity and quality of results; Supplier companies expect profitability and improved clinical outcomes; And regulatory agencies hope to reduce risks and improve patient safety. In all these aspects, the participation of biomedical engineers and clinical engineers is crucial especially when it comes to improving the technological environment of hospitals, as this task implies design, research, and definitely better quality of health services.

7.3 DESIGNING ALTERNATIVES FOR BETTER TECHNOLOGY IN HOSPITALS

One of the activities where we see that biomedical engineers and clinical engineers are little involved is in the development of clinical procedures, clinical practice guidelines (CPG), and clinical trials. It is true that these activities were once subjects very typical of health professionals, but it is also true that each time the technological component is present and entails considering the engineers as part of the working group. In this section we want to leave some notes in order to motivate the engineers in this important task and at the same time emphasize once again that the current organization of the hospitals is the biggest restriction for the accomplishment of these functions by the engineers.

As part of the program to improve health services based on the needs of the population, the development and adaptation of CPG requires additional resources and experience that can only be found in some health facilities, such as specialized hospitals or head of a network health. CPGs can change the way healthcare is delivered, improve patient outcomes and quality of service, ensure the efficient use of health resources, including technology, and enable the development of standards to evaluate clinical practice [6]. Usually a national entity is in charge of the CPG development, and the participation of the hospital is through its specialists, a work plan is elaborated where the CPG development is selected and prioritized according to different factors such as mortality and morbidity, variation in results or poor performance compared to other services. Likewise, the reporting of adverse events may generate the need for CPG. Also, new evidence or research may be the justification for developing a new CPG or adopting an existing one. It is valid to ask some questions before elaborating a CPG [7], for example: why you are elaborating a CPG? Why is it being elaborated now? Someone else has already elaborated? Is there available support of human and material resources for the elaboration of CPG? Is this a topic in which there is frequent litigation or is of high complication for the patient?

Once a topic is selected for guideline development, the steering committee will need to recommend whether the need is to [8]

- adopt a guideline
  - use existing guideline in current form
  - no changes apart from translation
  - still cost incurred in translation, implementation and review
- adapt a guideline
  - consider which recommendations from an existing guideline can be applied to the country or region
  - may require review of evidence
  - utilizes a formal process and has cost implication but a shortened time scale
- new Development
  - no guideline currently exists or is fit for adaptation to local practice
  - a new development required

Once the preparation of the CPG has been established, the Guide Development Group should be established, for which multidisciplinary members should be selected and this group will be different for each guide. This is where we want to emphasize that the participation of the biomedical engineer or clinical engineer will be very useful, given the technological components involved in any clinical procedure, and it is not only medical equipment, which already implies some complexity considering for example the ranges, precisions, accuracy, powers, and failure mode, but also the supporting technologies, such as environmental conditions, temperature, humidity, and ventilation, as well as information systems involved, data processing, signals and images, or systems for electrical protection and interference between medical devices, but also ensure the correct use of the physical variables for the appropriate effect on the tissues or the physiology of the organs, as well as the revision of the sequence of processes in relation to the norms and regulations, and regarding technologies of protection and asepsis they can also be better evaluated. In addition, in many cases, hospitals have already begun to use robotic or autonomous devices that require a previous mathematical analysis or use of models and simulations to design the sequence to be followed or to make specific configurations according to the conditions of each patient. At this point Human-Factor Engineering [9] is one of the great contributions that a clinical engineer or a biomedical engineer can apply to benefit an appropriate technological environment in a hospital.

Thus, of course, the preparation of CPG will require the necessary resources, such as specialized staff, according to project scope 10 or more members, time of participants, information technologies to facilitate information search, access to evidence, reduce number of face-to-face meetings. It is also required to cover the costs of implementation, audits, reviews, and updates, in addition to the required logistical support. In the case of CPG adaptation, the task is greatly reduced, both in terms of the personnel required and the necessary resources. The scope of the CPG is the first aspect to be addressed since the first meeting of the working group, it describes the purpose of the guide and identifies the areas to be covered and what will not be covered, details of the purpose of the patient group and professional group, clinical conditions, questions are added that the CPG should answer and contains a comprehensive bibliographic review that will guide the actions to be taken [7].

In the case of clinical trials, the objective is to evaluate the efficacy and safety of medical products and clinical procedures in humans, so new or improved medical treatments can be applied in medical practice. The three key points for the design of clinical trials are [10] controls, randomization, and blinding. Since in the last 30 years the health publications have increased three times, but clinical trials increased 20 times in the same period, then it is expected that in the next years there will be a high activity and therefore it will be appropriate to summon engineers and others professional profiles for this task. A first aspect to be solved in the design of clinical trials is that the estimation of efficacy or safety will never be absolutely conclusive since we start from the study in a sample and not in the whole population, reason why it is always an approximation to validate, with the purpose of reducing the probability of false results, false negatives and false positives, and this will be dependent on multiple factors such as the objectives of the study, therapeutic area, comparison between treatments and phase of clinical trial. Here the basic problem is to establish the characteristics of the random sample that represents the population studied, and we rarely know how representative is a sample of the real world. Thus, we must decide what level of risk we are willing to assume and rationally justify, since accepting a certain false negative, for example would be very costly for patients and society due to the losses involved from wrong treatment.

The design of clinical trials follows several criteria and methods that are described below, among the most common ones:

- Parallel group design, where participants are randomly grouped into one of two treatment groups. One of the groups is given the object of the test and the other is given a placebo or the standard procedure.

- Another type of test design is the cross-over, in which one has a group that is given sequentially the new treatment and then the placebo or the standard treatment, and the other group is given the reverse sequence. This method has the advantage of eliminating the individual difference of the participant with respect to the effect of the complete treatment. Cross-over design is also more sensitive to dropout during the course of the trial because participants act as a control group and also as active treatment participants.
- The clinical trial called the Open-Label Trial is less common, it is when both research groups and participants know which treatment will be administered, only that the group of participants will randomly receive one of the two treatments. Sometimes a trial has more than two treatment groups, it is used when different doses will be compared to each other.
- One type of increasingly popular trial is the so-called Adaptive Clinical Trial. Examples of adaptive trial designs include setting aside a treatment group, modifying sample size, balancing treatment assignments by adaptive randomization, or simply stopping an early trial because of success or failure found. In a standard trial, safety and efficacy data are collected and reviewed by a safety and data monitoring committee during the scheduled interim analyses. However, apart from stopping a trial for security reasons, very little can be done in response to these data. Often, a completely new trial should be designed to further investigate the key findings of the trial.
- Selection of control group is one of the most critical aspects. The control group experience tells us what would have happened to participants if they had not received the test treatment—or if they had received a different treatment known to be effective. A control group is chosen from the same population as the test group and treated in a defined way as part of the same trial studying the test treatment. Test and control groups should be similar to the initiation of the trial on variables that could influence outcome, except for the trial treatment. Otherwise, bias may be introduced into the trial. The type of control can be placebo, no treatment, different dose or regimen of the trial test, or the standard treatment [10].

Clinical trials, in their different varieties and approaches, need to be designed in detail and are not alien to the use of mathematical applications, statistics, and experimental management of the science and engineering laboratories. We believe that the professional profiles of biomedical engineers and clinical engineers will be of great help to researchers in the area of health. Table 7.1 shows a summary of the study types according to the objectives to be achieved in the clinical trials.

Table 7.1 An Approach to Classifying Clinical Studies According to Objective [11]

Type of Study	Objective of Study	Study Examples
Human Pharmacology	• Assess tolerance • Define/describe Pharmacokinetics and Pharmacodynamics • Explore drug metabolism and drug interactions • Estimate activity	• Dose-tolerance studies • Single and multiple dose Pharmacokinetics and/or Pharmacodynamics studies • Drug interaction studies
Therapeutic Exploratory	• Explore use for the targeted indication • Estimate dosage for subsequent studies • Provide basis for confirmatory study design, endpoints, methodologies	• Earliest trials of relatively short duration in well- defined narrow patient populations, using surrogate or pharmacological endpoints or clinical measures • Dose–response exploration studies
Therapeutic Confirmatory	• Demonstrate/confirm efficacy • Establish safety profile • Provide an adequate basis for assessing the benefit/risk relationship to support licensing • Establish dose–response relationship	• Adequate, and well controlled studies to establish efficacy • Randomized parallel dose–response studies • Clinical safety studies • Studies of mortality/morbidity outcomes • Large simple trials • Comparative studies
Therapeutic Use	• Refine understanding of benefit/risk relationship in general or special populations and/or environments • Identify less common adverse reactions • Refine dosing recommendation	• Comparative effectiveness studies • Studies of mortality/morbidity outcomes • Studies of additional endpoints • Large simple trials • Pharmaco-economic studies

Depending on the type of technology (clinical, supportive, preventive, protective, promotional, or environmental), some types of study and methods will be more appropriate, naturally have more pharmacological studies; however, with the greatest boom in medical devices and therapies many studies are still required to establish the most effective doses and configurations.

Whenever an investigation is considering the involvement of human beings, we must keep in mind the functions of the Ethics Committee for Research or Human Subjects Research Committee. Its mandate is to ensure the ethical commitment of researchers, as well as to certify and supervise that research carried out or promoted by the hospital complies with the ethical principles of research. Ethical principles can be described in various ways, here a version of the principles of research with human beings [12]:

1. *Respect for people*
   This principle demands the recognition of the autonomy of the people and the protection of those whose autonomy is somehow diminished. Respect for this principle will mean that persons who are

subject to research have the appropriate information and that they voluntarily participate in research and that they may also withdraw if they so decide; As well as full respect for their fundamental rights, in particular if they are in a situation of special vulnerability, be it age, illness, mental decline, or because they are subject to economic, cultural, or other circumstances that affect their autonomy.

2. *Beneficence and nonmaleficence*

   It is the duty of the researcher to ensure the welfare of the persons involved in the research. The researcher must ensure that his/her conduct cause not harm to researched nor researchers nor third parties. Likewise, the researcher must strive to reduce or compensate for possible adverse effects and to maximize the benefits of research.

3. *Principle of Caution or Precautionary*

   Where the investigation entails a danger of harm or disorder that is serious or irreversible, the lack of absolute certainty about such danger should not be used as a reason for delaying the adoption of effective measures to prevent it. This requires, to the fullest extent possible, a comprehensive assessment, the determination of the degree of uncertainty and the assessment of risks. Researchers are obliged to ensure compliance with the precautionary principle both in the researches they carry out and in the dissemination of results, recognizing the right of society to know the long-term benefits and risks arising from the acquisition of new scientific knowledge or new research methods.

4. *Justice*

   The researcher must exercise reasonable judgment and take the necessary precautions to ensure that his/her biases, and limitations of his/her abilities and knowledge, do not give rise to or tolerate unfair practices. A benefit should not be denied to a person who is entitled to it, without any reasonable reason, or unduly impose a burden. It is recognized that equity and justice grant the right to access to their results to all persons involved in the research. The confidentiality of a researcher with respect to the personal information of the subjects participating in his/her studies, the professional secrecy previously committed to any of the parties, or the common good of the society, are limits to this right. Likewise, the researcher must treat equitably those who participate in the processes, procedures, and services associated with the research.

5. *Scientific Integrity*

   This principle demands honest and truthful action in the collection, use, and conservation of the data that serve as the basis for a research, as well as in the analysis and communication of its results. The integrity or rightness should govern not only the scientific activity of the researcher, but also include the assessment of the origin of the funds and the procedures used to obtain them, in addition to extending to their teaching activities and their professional practice. The integrity of the researcher is especially relevant when depending on the ethical standards of his/her profession, potential damages, risks, and benefits that may affect those involved in an investigation are evaluated and declared. This includes the need to declare conflicts of interest that could affect the course of a study or the communication of its results.

6. *Responsibility*

   The researcher must be aware of his/her scientific and professional responsibility to society. In particular, it is the researcher's personal duty and responsibility to carefully consider the consequences that the realization and dissemination of his/her research imply for all participants in this and for society in general. This duty and responsibility can't be delegated to other people. Neither the act of delegating nor the act of receiving instructions frees of responsibility. It is up to the person conducting the research to ensure that the persons on the team are qualified to perform their assigned duties and are responsible for compliance with the principles contained in these regulations. Ignorance of these ethical and normative principles does not exempt the researcher or his/her team from responsibility for the consequences of their researches.

In addition, for purposes of research with human beings, the following are duties of researchers:

1. Provide information to the participants on the objectives, nature of the research, the uses to be made of the information collected, the possible risks and benefits, and any doubts that the participant would like to resolve regarding the research.
2. Ensure the confidentiality of the information and knowledge provided by participants using appropriate procedures. According to the research purposes, the researcher may consider the option of anonymity of participants rather than confidentiality. When requested by the participants or when the researcher deems it necessary, the authorship and the contribution of the participants in the production of the knowledge must be made explicit or the anonymity of the participants must be assured.
3. Respect the freedom and autonomy of the subjects to participate in the research or to withdraw of that if they so decide.
4. In the case of researchers with participants in an interdict or in case of minors, the informed consent of the legally authorized person must be obtained.

5. When researches include underage participants from the age of 12 years, they should be asked for informed consent. The informed consent of the children does not exclude obtaining the informed consent of the persons legally responsible for them, except in those cases in which the problems of study so require.
6. Researchers should not conduct researches involving deception techniques to participants, except in circumstances where their use is fully justified and the benefit of the research far outweighs the possible involvement of such participants.
7. Comply with the commitments made with the participants in their research and with the institutions that have provided some kind of collaboration with the study.
8. Perform the return of global results of the study whenever possible, or failing that, provide access to those.
9. Disseminate the findings of their research to the attached community and to society as a whole in order to contribute to their knowledge and development.

The role that engineers can assume in relation to ethical principles for research is today more than ever of great importance due to their training and knowledge of the technological issues involved in research, so we think this approach is one of the reasons that may motivate engineers to work in a hospital.

7.4 TRANSFER TECHNOLOGY, STARTUPS, AND SCIENTIFIC AND TECHNOLOGY PARK

There is a certain potential where hospitals and hospital networks are increasingly involved, and that is yes, they have the capacity to generate new or better technology in health, then it can acquire an economic value in the market, or at least promote business initiatives that reward the hospital in an economic way. It is true that the main role of the hospital, which is to provide quality health services, should not be lost, but in this effort to improve and research, to solve problems, and develop alternatives, that may well have an economic value or a social value for others. And so many health systems have begun to work in this capacity and make the operation of the hospital more viable. We believe that the participation of the engineers can guide these initiatives and at the same time contribute to the primary purpose that is to provide health to the population.

In this context, a first aspect to consider is the processes of technology transfer. Technology transfer is the process of transferring scientific findings from one organization to another for the purpose of further development and commercialization. It is shaped by the environment with respect to local capacity, ownership of intellectual property, availability of finance, and other factors. In May 2008 and 2009 the World Health Assembly adopted Resolutions WHA61.21 and WHA62.16 on the Global Strategy and Plan of Action on Public Health, Innovation and Intellectual Property (GSPA-PHI) [13]. The plan incorporates eight elements: Prioritize research and development (R&D) needs; Promote R&D; Build and improve innovative capacity; Transfer of technology; Application and management of intellectual property to promote innovation and promote public health; Improve delivery and access; Promote sustainable financing mechanisms and Establish and monitor reporting systems. Technology transfer and local production are often motivated by cheaper production costs and easier penetration into emerging markets.

We must remember that health technology involves different types of technology, even when we find more reports on pharmacological experiences of technology transfer, each type of technology has a potential in the market, and hospital engineers must analyze the context of their locality for find the best opportunities. Table 7.2 shows some recommendations for improving technology transfer capacity in the context of developing countries taking example for diagnostic technologies.

The experiences of technology transfer from hospitals are diverse in the health sector. Here we present a couple of examples:

1. The National Cancer Institute's Technology Transfer Center (TTC) [14] facilitates partnerships between the National Institute of Health (NIH) research laboratories and external partners. With specialized teams, TTC guides the interactions of our partners from the point of discovery to patenting, from invention development to licensing. We play a key role in helping to accelerate development of cutting-edge research by connecting our partners to NIH's world-class researchers, facilities, and knowledge. How We Can Help? If you are
   a. a *pharmaceutical or biotechnology company representative* interested in solving a specific technical issue or developing your business and building strategic alliances, we invite you to search for available technologies and/or contact us to locate available technologies; or
   b. a *university researcher* interested finding a collaborator or research materials or a *nonprofit organization* interested in finding treatment options for patients or to promote the development of NCI technologies into novel therapies; or
   c. an *NIH Investigator* interested in patenting your discovery or in need of resources to develop your technology that are not available at

Table 7.2 Recommendations for Enhanced Technology Transfer and Local Production for Example to Diagnostic Technology in Developing Countries [13]

Promote	Support	Develop
Advocate the importance and value of diagnostics to the pharmaceutical industry, and national and international Stakeholders	Provide guidance on required test specifications for developing countries for test developers	Develop new business models and approaches to financing and marketing of diagnostics
Analyze the developing world market and provide data to test developers, potential investors, and local stakeholders	Provide critical pathway for successful technology transfer and examples of best practice	Explore novel initiatives by which to share and exploit intellectual property
Enhance capacity within developing countries to adopt new diagnostic and manufacturing technology	Provide advice and training on protecting intellectual property. Collate and distribute information on patents and where they apply	Establish a global association for manufacturers of diagnostic tests
Recognize excellence in diagnostic expertise and establish a professional career pathway in diagnostics R&D	Support training to enhance capacity for Good Laboratory Practice GLP/ ISO manufacturing in developing countries	Establish a global diagnostics forum for information sharing, enable debate, and encourage collaboration and harmonization
	Support capacity building to increase the number of stringent regulatory authorities in developing countries	Develop a health technology assessment model for countries to determine whether new technologies address their public health needs

NIH, NCI TTC can help you find companies to co-develop or license your technology.

2. Technological Transfer of Knowledge in Pulmonology (Neumology Service of University Hospital Germans Trias i Pujol, Badalona, Barcelona, Spain) [15]. In Spain, some research groups have been able to reach the preindustrial production and marketing phase. The following is a description of some of the projects that have transferred or are in the process of transferring technology. The group at the Hospital del Mar de Barcelona (IMIM) has developed a training device for the respiratory muscles that uses resistive loads (ORYGEN DUAL VALVE, Forumed, Spain). The device is the result of more than 15 years of research in respiratory muscle function. Electrical impedance tomography (EIT) has been successfully introduced by the group at Hospital de San Pablo in Barcelona (TIEsys-4, Barcelona, Spain). EIT offers thoracic images that give quantitative parameters of the pulmonary ventilation. EIT does not require ionizing radiation and gives results similar to those offered by perfusion gammagraphy. The group constituted by the Hospital del Río Ortega in Valladolid and the Hospital Clínico de Santiago de Compostela has studied

pulse-oximeter signals and heart rate for 10 years. By means of spectral and central tendency analyses, they have been able to discriminate patients affected by SAHS. The "Disavoz" Project by the group at the Hospital Arnau de Vilanova in Lérida is trying to identify patients affected by AHS by means of voice analysis based on previous studies (SEPAR grant, 2009). The analysis of snoring using the "Snorizer" system (in process of being commercialized by Sibel SA, IBEC, Hospital Universitari Germans Trias i Pujol, Badalona, Barcelona, Spain) can differentiate between the different degrees of severity in patients with SAHS. At the same time, important advances are being made in the field of telemedicine applied to home CPAP titration.

The range of possibilities of technology transfer is very wide. Naturally the research is carried out with associated entities such as universities or research institutes or in conjunction with companies, so the diversity of technological alternatives is even greater. Table 7.3 shows examples of technology transfer promoted by the World Health Organization (WHO).

The transfer of technology is only a first step on the road to the development of a biomedical industry. In addition to the administrative processes of creation,

Table 7.3 Examples of Technology Transfer on Health Sector [13]

Technology Transfer	Example
Transfer of manufacturing by multinational company. In this model manufacturing technology is transferred to a subsidiary site in a developing country	A United States based company, Orasure Technologies Inc., has contracted a company in Thailand to assemble its OraQuick HIV device
Diagnostic R&D by multinational company at the developing country site. In this model product development but not manufacture is undertaken at the subsidiary site	Development of molecular test for mycobacteria in South African laboratories by Roche Diagnostics. The test was designed to meet the needs of South Africa, which has a significant number of cases of pulmonary infection with mycobacteria other than TB
Partnership of local small- or medium-sized enterprise or nonprofit-making organization with multinational company for technology transfer and sharing of knowledge	Production of rapid tests for priority diseases by the Immunobiological Technology Institute (Fiocruz) in Brazil in partnership with Chembio Diagnostic Systems, Inc.
Partnership of commercial test developer/ manufacturer with international nonprofit-making organization	India: various companies have collaborated with PATH for lateral flow point-of-care technology
Partnership of nonprofit-making organization with nonprofit-making international organization	PATH or FIND partnership with reference laboratories to evaluate tests.
Procurement of technology by developing country test developer to enable independent development and manufacture of a novel test	South Africa: NHLS have developed nonprofit-making in-house tests
Development and manufacture of diagnostic test using indigenous knowhow and technology and knowledge obtained from published literature	South Africa: Vision Biotech/Alere Healthcare Pty1—rapid tests for HIV, malaria, and human chorionic gonadotrophin (HCG)

registration, and regulation of companies, we are interested in showing the promotion of new companies or Startup, which requires the support of business incubators or at least the support of specialized offices, and on the other hand, the strong regional approach of the company–university–government association, or, in this case, hospital–university–government, for the development of an industry through science and technology parks.

Why Start a Company? When launching a new company, there are a number of factors to take into consideration. The reality is that only some inventions may be suitable for the creation of a startup company. Innovations may progress more quickly in a focused startup than in an academic lab or a large company. Along with the invention team, a specialized office helps in analyzing several factors to determine whether a startup is the most appropriate path to commercialization [16]:

- Demand: Potential of the core technology to provide a solid platform for multiple markets or product opportunities.
- Competition: Identification of other companies that offer similar solutions.
- Licensing: Likelihood of interest from existing companies in licensing the technology.
- Funding: Availability of capital to build and grow the business, together with the interest, capabilities, and track record of likely investors.
- Commitment: Level of commitment and involvement of the inventors.
- Support: Presence of a true business champion for both the technology and the new venture.
- Management: Experience, passion, and drive of the startup's executive team.

At least three aspects are considered at the time of proposing the creation of a Startup company:

1. Business Model Innovation: A plan for the successful operation of a business, identifying sources of revenue, the intended customer base, products, and details of financing. Is where two or more elements of a business model are modified to deliver better value in a new way. The business model Canvas [17] is actually very useful, this is a complex concept referring to the complete design of the activities of a company, take into account the "global picture," and defining all the parameters and characteristics of the company and its relations with its environment. The business model Canvas is a tool gathering all components of the strategy and the operations, and helping the manager/CEO to plan his/her business.
2. Lean Startup: Is a process for turning ideas into commercial ventures. Its premise is that startups begin with a series of untested hypotheses.

They succeed by getting out of the building, testing those hypotheses, and learning by iterating and refining minimal viable products in front of potential customers.
3. Continuous Improvement: Constantly seeking and gaining feedback from customers to adjust business processes in order to deliver better value

In the way of developing technology in health with the participation of hospitals, then transferring it and making it reach to the population, through startup companies, we have the integrative approach at the level of a state or a complete region that is the creation of science and technology parks. Clearly this is an innovative approach for hospitals, but we see that it must be considered because of successful strategies in several countries. The World Technopolis Association (WTA) describes the importance of science and technology parks in the following terms [18]:

7.4.1 An Era of Knowledge-Based Societies

The development of science and technology leads directly to the welfare and prosperity of mankind. Only countries, cities, and enterprises that excel in cutting-edge science and technology will be able to take leading roles in this new era. In particular, regional competitiveness is crucial to national competitiveness in this age of localization and globalization. Therefore, local governments perform a variety of developmental strategies to enhance regional competitiveness in order to take full advantage of regional potential for locally generated growth and innovation.

7.4.2 Technopolis to Boost Regional Economic Growth and Innovation

Technopolis development based on science and technology parks can boost economic development and increase innovative capacity in a region. It aims to enhance innovative capacity and foster strategic industry through the interaction and close linkage among governments, research institutes, HEIs, and high-tech industries for greater regional and national competitiveness. It also plays an important role in complementing regional business activities as a catalyst for change by both promoting the establishment of new businesses and furthering the growth of existing businesses.

7.4.3 Global Networks for Mutual Cooperation Among Technopolises

By competing against each other, Technopolises enhance their competitiveness in the era of knowledge-based society. The creation of collaborative partnerships and the promotion of knowledge and technology exchange among Technopolises can be extremely useful to boosting global competitiveness. In particular, successful experiences from world science cities can upgrade hi-tech capability through closer cooperation between local governments, HEIs, R&D centers, and hi-tech industries in a region.

We believe that, over time, hospitals, through a holistic and appropriate management of their technology and applied research, will be able to solve their technological problems for the provision of quality health services. In this effort, the participation of biomedical engineers and clinical engineers in hospitals will be crucial. However, the ability to solve problems and with a creative approach, hospitals can become part of a network generating new and improved technologies for health. Such a scenario will be very attractive for the active participation of biomedical engineers, clinical engineers, and all other health professionals committed to providing better health services to the population.

References

[1] WHO. Health research methodology: a Guide for training in research methods. 2nd edition World Health Organization—Regional Office for the Western Pacific Manila; 2001.

[2] Salustri FA. Is it time to separate Applied Science and Engineering? In: Proceedings of 1st annual conference of the Canadian design engineering network, Montreal; 2004.

[3] West MA, Farr JL. Innovation at work. In: West MA, Farr JL, editors. Innovation and creativity at work: psychological and organizational strategies. Chichester, England: Wiley; 1990. p. 3–13.

[4] Omachonu VK, Einspruch NG. Innovation in healthcare delivery systems: a conceptual framework. Innov J: Public Sect Innov J 2010;15(1) Article 2.

[5] Jusko J. Innovation: measuring success. Businesses link innovation to performance metrics. IndustryWeek. April 14, 2005. <http://www.industryweek.com/product-development/innovation-measuring-success>; April 15, 2017.

[6] NICE. The guidelines manual. National Institute for Health and Clinical Excellence. NHS, London; 2009.

[7] Scottish Intercollegiate Guidelines Network (SIGN). SIGN 50: a guideline developer's handbook. Edinburgh, Revised edition published; 2015.

[8] Shillito J, Rathod M. Guideline development and adaptation. Royal College of Obstetricians & Gynaecologists Global Health Toolkit N° 5; 2014.

[9] Wickens CD, Gordon S, Liu Y, Lee J. Introduction to human factors engineering. 2nd edition, New York: Pearson; 2003.

[10] Karlberg J, Speers M. Reviewing clinical trials: a guide for the ethics committee, 2010. Washington, DC, USA: Association for the Accreditation of Human Research Protection Programs, Inc.; Hong Kong SAR, PR China: Clinical Trials Centre, The University of Hong Kong.

[11] EMEA. General considerations for clinical trials. London: European Medicines Agency. ICH Topic E 8; 1998.

[12] PUCP. Bylaw of the Research Ethics Committee. Pontifical Catholic University of Peru; 2016<http://cdn02.pucp.education/investigacion/2016/10/14160435/Reglamento-2.pdf>. April 17, 2017.

[13] WHO. Increasing access to diagnostics through technology transfer and local production. Geneva: World Health Organization; 2011.

[14] The National Cancer Institute's Technology Transfer Center (TTC). <https://techtransfer.cancer.gov/abouttc>.

[15] Fiz JA, Morera J. Technological transfer of knowledge in pulmonology. Arch Bronconeumol 2012;48(5):141–3, http://dx.doi.org/10.1016/j.arbr.2011.11.013.

[16] Harvard University. Startup guide: an entrepreneur's guide for Harvard University faculty, graduate students, and postdoctoral fellows. Office of Technology Development. 2011, Cambridge.

[17] Osterwalder A, Pigneur Y. Business model generation. Amsterdam: Alexander Osterwalder & Yves Pigneur; 2009. ISBN: 978-2-8399-0580-0.

[18] WTA. Global network leading to sustainable technopolis development. World Technopolis Association, Daejeon.<www.wtanet.org>.

CHAPTER 8

Improvement Healthcare Projects: Meeting Healthcare and Technology Challenges

Shaping approaches to improve the Quality of healthcare solutions

8.1 HEALTHCARE CHALLENGES

Safety, affordability, access, new infectious diseases and prevalent chronic conditions, the complexity of the healthcare environment, the complexity related to the healthcare organization itself, are some challenges of today's healthcare.

Over time public health's challenges have changed: rapid identification, triage, and treatment of acutely sick and injured patients in cases like influenza (Influenza pandemic plan. The role of WHO and Guidelines for National and Regional Planning, World Health Organization, Communicable Disease Surveillance and Response, Geneva, 1999.) and severe acute respiratory syndrome—SARS (Severe acute respiratory syndrome (SARS): Status of the outbreak & lessons for the immediate future, World Health Organization (WHO), Communicable Disease Surveillance and Response, Geneva, 2003.); viral diseases like chikungunya (World Health Organization, 2016. http://www.who.int/mediacentre/factsheets/fs327/en/); the assessment, communication, and mitigation of public health risk; and the provision of assistance to state and local health officials to quickly reestablish healthcare delivery systems and public health infrastructures (The White House, 2006); the management of health effects of "El Niño" climate conditions (El Niño and Health, Global Overview, World Health Organization, Geneva, 2016.), urgent and effective public health services for injured people due to terrorist attacks (Vandentorren S, et al. Syndromic surveillance during the Paris terrorist attacks. The Lancet 387(10021);846–7.) are some of the evidences; the increasing impact on medicine due to the utilization of health technology: artificial organs, robotic prosthetic limbs, etc., which define a complex context for healthcare.

Healthcare Technology Management Systems. DOI: http://dx.doi.org/10.1016/B978-0-12-811431-5.00008-4
© 2017 Elsevier Inc. All rights reserved.

1. Absence of a clear, realistic vision of the public health system
2. Lack of organizational leadership
3. Inexistent strategic alignment of resources
4. Poor level of quality to implement the process of change required

FIGURE 8.1
Some obstacles for the biodefense role of public health. *Salinsky E, Gursky E. The case for transforming governmental public health. Health Affairs 25(4):1017–2018. Available at: <http://content.healthaffairs.org/content/25/4/1017.full>; 2006.*

These challenges determine a relevant change in the way we understand health systems, Fig. 8.1 illustrates some obstacles in the case of the biodefense role of public health for example; in this regard Salinsky [1] states, "A transformed public health system is needed to address the demands of emergency preparedness and health protection. . . .The future public health system cannot afford to be dictated by outmoded tools, unworkable structures, and outdated staffing models."

8.2 TRENDS AND EMERGING TECHNOLOGY

Mega Trends are global, sustained, and macro-economic forces of development that impact business, economy, society, cultures, and personal lives thereby defining our future world and its increasing pace of change.

On the top of the "Top 10 Mega Trends to 2020," eight have significant emerging technology components. Emerging technologies will be the dominant driver of disruptive change for the future [2]. Healthcare systems despite the level of their economic sector should be aligned to the change. See Fig. 8.2A, B:

The Mega Trend number 8: "Health, Wellness & Wellbeing" facilitates the understanding about (1) the increasing level of complexity of the health organization; and (2) the value of the multidisciplinary workforce (physicians, nurses, engineers, managers, biologists, chemists, etc.) capacity to work as a team to be aligned to the current and expected challenges. See Fig. 8.3.

8.3 FACING THE CHANGE IN HEALTH SYSTEMS

The Healthcare trends experimented by developed and developing countries define a relevant change for health systems. An interesting contribution from

(A)

1. **Urbanization** "City as a Customer" Mega Cities, Mega Regions, Mega Corridors, and Mega Slums. Cities, and Not Countries, Will Drive Wealth Creation.
2. **Smart** is the New Green Smart Cities, Smart Technology, Smart Infrastructure, Smart Energy, Smart Mobility, Smart Buildings, Smart Clouds, Smart Materials, etc.
3. **Social Trends** Gen Y, Geo Socialization, "She-economy," Ageing Population, Reverse Brain Drain, etc.
4. **Economic Trends** Power Shift from West to East. Beyond BRIC–the next game changers based on GDP by 2025 = Indonesia $4.8trn, Mexico $2.8 trn., Turkey $2.4trn, Poland $1.2trn, etc. Africa–major future source of the worlds resources.
5. **Connectivity & Convergence** 80 Billion Connected Devices By 2020. 5 connected devices for every user by 2020. 5 billion internet users by 2020 (over 60% of world's population). Connectivity will Accelerate Convergence.

(B)

6. **Innovating to Zero** Mindset change in how to think/plan innovation ie. Carbon neutral cities, zero email.
7. **New Business Models** "Value for Many" will replace "Value for money," i.e., Freemium, Group Buying.
8. **Health, Wellness and Wellbeing** Power shifts to the patient. Patient Centric Connected Health. Connected Health Driving New Access to Care Solutions. New treatments and patterns of care ie E-Health/M-Health, Gene Therapy, Health Kiosks, Tissue Engineering, Healthcare Tourism, Cybernetics, Noninvasive Surgery.
9. **Home-centering** Insourcing jobs, the home as the center for work not offices, home as distribution point, more services will be targeted for home delivery not centrally, i.e., health care moving away from hospitals to be home focused.
10. **Tech Vision 2020** Expect dramatic advances in 9 technology areas, in priority order; Sustainable Energy, Clean & Green Environment, Health & Wellness, Information & Communication, Materials & Coatings, Medical Devices & Imaging Tech, Microelectronics, Sensors & Control, Advanced Manufacturing and Automation.

FIGURE 8.2
(A, B) Top 10 Mega Trends to 2010. *Frost & Sullivan Manufacturing Leadership Council. Top 10 Mega Trends, New York, 2012.*

Hayes (Emerging business & technology trends, ISACA, 2014.) states that the management of emerging technologies will require transformational changes in some or all of the seven enablers in the healthcare organizations see Fig. 8.4.

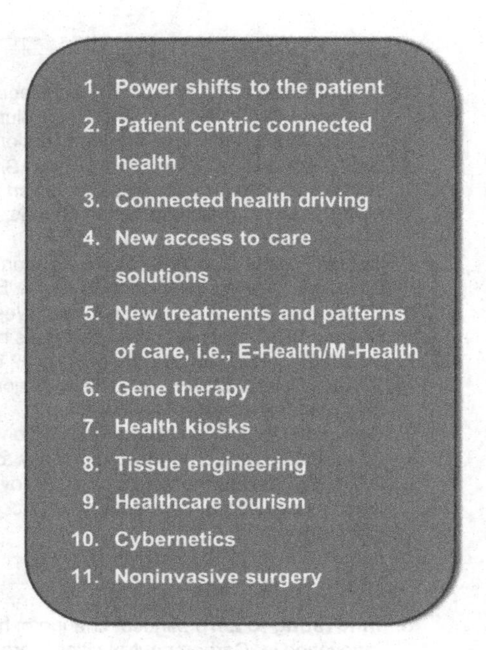

1. Power shifts to the patient
2. Patient centric connected health
3. Connected health driving
4. New access to care solutions
5. New treatments and patterns of care, i.e., E-Health/M-Health
6. Gene therapy
7. Health kiosks
8. Tissue engineering
9. Healthcare tourism
10. Cybernetics
11. Noninvasive surgery

FIGURE 8.3
Mega Trend 8: "Health, Wellness & Wellbeing." *Frost & Sullivan Manufacturing Leadership Council. Top 10 Mega Trends, New York, 2012.*

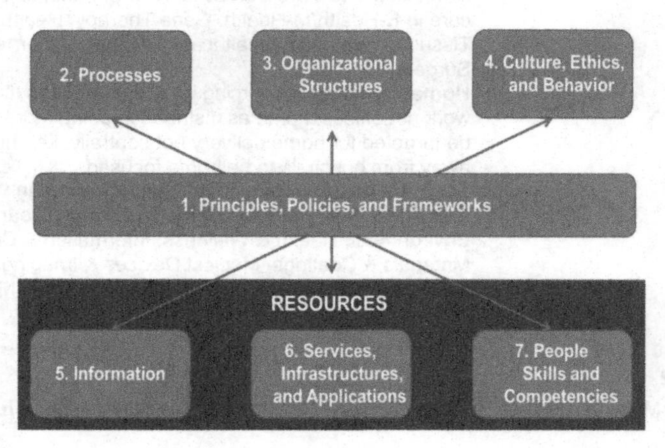

FIGURE 8.4
Managing the change: seven enablers to identify organizational capability areas. *Hayes T. ISACA, 2014.*

The consideration of the enablers contributes to the understanding that the management of technology from an effective and consistent perspective dismisses the link of health technology exclusively with operational and/or technician areas.

8.3.1 Systems Engineering and Healthcare

Systems Engineering (SE) is a systems approach which focuses on developing solutions aligned to (1) economics; (2) technology; (3) social dynamics; and (4) healthcare policy. This perspective is consistent to high, medium, or small levels of economy. A systems approach involves thinking holistically and work with transdisciplinary teams to develop solutions [3].

The Institute of Medicine—IOM [4] and the National Academy of Engineering—NAE [5] recommended and advocated the widespread application of SE tools to improve healthcare delivery [6]. Despite the differences between the levels of the economic sector, healthcare environments have in common: management, project planning, inventory, logistics, facilities design, process flow analysis, resource synchronization, etc. This framework aims on improving the analysis and results expected.

Kopach-Konrad [7] emphasizes that the application of SE requires medical professionals and managers understand and appreciate the power that SE concepts and tools can bring to (1) redesigning, and (2) improving healthcare environments and practices.

8.3.1.1 Definitions

SE focuses on the design, control, and orchestration of system activities to meet performance objectives. A *System* is a set of possibly diverse entities (patients, nurses, physicians, etc.), each performing some set of functions. The interaction of these entities as they perform their various functions gives rise to a global *System Behavior*.

8.3.1.2 Healthcare Improvement Project Management: SE Model

Fig. 8.5 illustrates the steps oriented to manage a healthcare improvement project according to SE model [7].

SE approach is pertinent to the growth, operation, and synchronization of many information-rich and technologically complex economic sectors, as described above health related to developed and developing countries is certainly an interesting sector to SE's application. The following is one of the examples:

8.3.1.3 Improving the Effectiveness and Efficiency of an Intensive Care Unit

Applied Physics Laboratory—APL and Johns Hopkins Medicine—JHM [8] applied successfully SE to achieve effectiveness and efficiency objectives in the intensive care unit—ICU (Project funded by the Johns Hopkins University Whiting School of Engineering Systems Institute—WSE-SI to study integration and interoperability opportunities and challenges in the ICU,

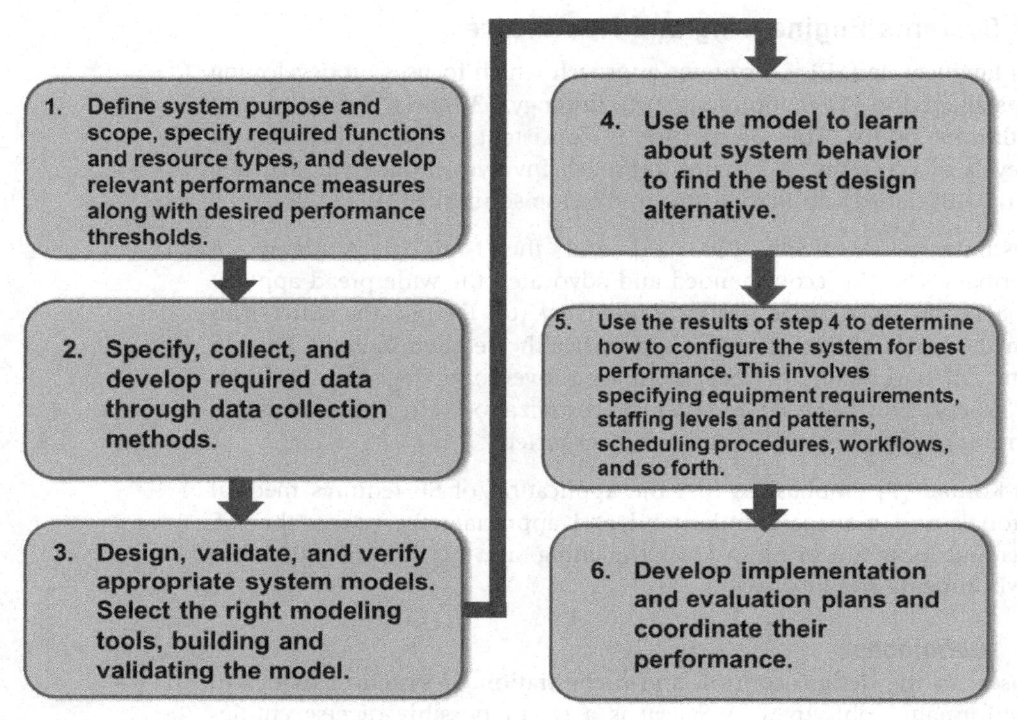

FIGURE 8.5

Steps to manage a healthcare improvement project. *Kopach-Konrad R, Lawley M, Criswell M, et al. Applying systems engineering principles in improving health care delivery. J Gen Intern Med 2007;22(Suppl. 3):431–7. doi: 10.1007/s 11606-007-0292-3.*

emphasizing the role of the patient and family in their own care within the ICU, USA, 2011–2012.). The objective was on identifying where and how integration and interoperability could improve clinical situational awareness and command and control.

The new ICU system has an information display system based on a common Integrated Clinical Picture—ICP user interface. ICP is designed for rapid, intuitive information integration, assimilation, and sense-making. The ICU system also provides the ability to control the state of clinical systems (infusion pumps, ventilators, and other medical devices) and nonclinical systems (lighting, heating, ventilation, television controls, etc.).

Fig. 8.6 presents the interactions analyzed by APL-JHM team in order to formulate a systems approach to innovations oriented to improve patient outcomes at the ICU.

The process included the development of measures of effectiveness and measures of performance that quantitatively and qualitatively provided guideposts to improve safety and quality in healthcare delivery. Below the "V-model," sequence of system development, test & evaluation, and fielding utilized by the team, see Fig. 8.7.

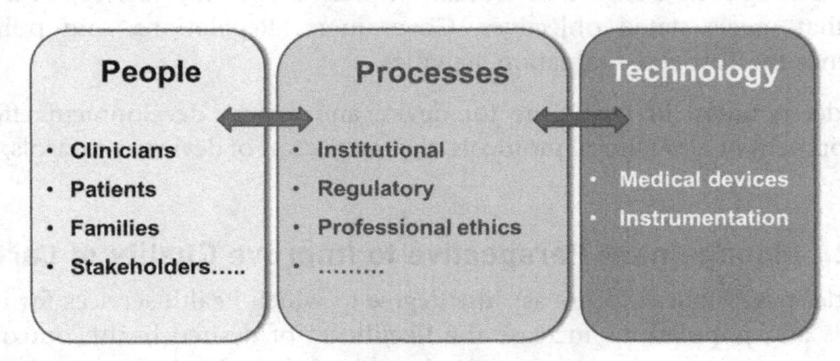

FIGURE 8.6

Analysis of interactions to improve patient outcomes. *Ravitz A, et al. Systems approach and systems engineering applied to healthcare: improving patient safety and healthcare delivery. Johns Hopkins APL Tech Dig 2013;354(31):4.*

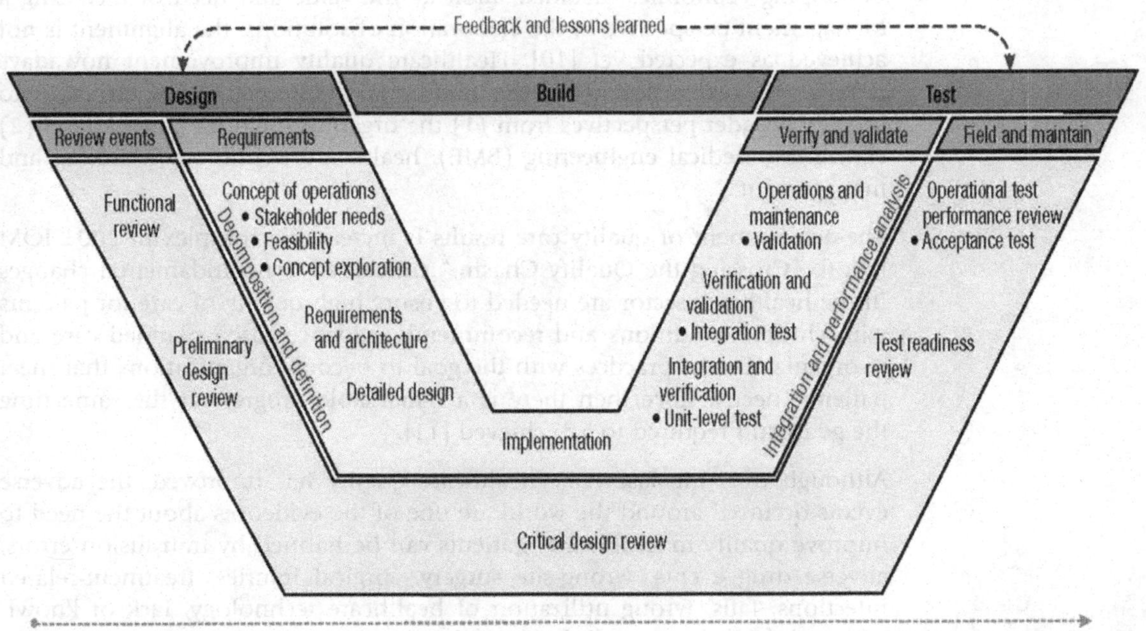

FIGURE 8.7

System development life cycle. *Ravitz A, et al. Systems approach and systems engineering applied to healthcare: improving patient safety and healthcare delivery. Johns Hopkins APL Tech Dig 2013;354(31):4.*

By using V-model health workforce team adheres to SE's best practice of maintaining a comprehensive set of technology, documents, execution of reviews and analyses. V-model that enables the system requirements may be traced through the design and evaluation phases, ensuring delivery of a system that meets stated objectives. Government, Regulations, and policies influence the Test and Evaluation activities.

V-model is useful in healthcare for device and system development efforts, development of new clinical protocols, the integration of devices, protocols, etc.

8.3.2 Management Perspective to Improve Quality of Care

IOM defines Quality of Care as: "the degree to which health services for individuals and populations increase the likelihood of desired health outcomes and are consistent with current professional knowledge" [9].

Some reasons for low quality in health services are inadequate provision of care, limited resources, and the rising costs by inappropriate care. In this regard, a relevant number of evidences around the world, developed and developing economies included, indicate the value and need of including a Management perspective in the elaboration of solutions, the alignment is not achieved as expected yet [10]. Healthcare quality improvement nowadays requires the expansion from the traditional interpretation of structure to include broader perspectives from (1) the organizational framework; and (2) clinical/biomedical engineering (BME), healthcare technology planning and management.

The achievement of quality care results is increasable complex in 2001 IOM report, "Crossing the Quality Chasm," concluded that fundamental changes in the healthcare sector are needed to ensure high quality of care for patients with chronic conditions and recommends evidence-based planned care and re-organization of practices with the goal to become organizations that meet patients' needs, since then there is a remarkable progress at the same time the goals still required to be achieved [11].

Although over the last years healthcare quality has improved, the adverse events occurred around the world are one of the evidences about the need to improve quality in healthcare. Patients can be harmed by transfusion errors, adverse drug events, wrong-site surgery, surgical injuries, treatment-related infections, falls, wrong utilization of healthcare technology, lack of knowledge/training related to clinical guidelines, etc.

A study of Classen et al. ("Global trigger tool" shows that adverse events in hospitals may be ten times greater than previously measured. Health affairs

2011;30(4):581–9.) determined that in US healthcare system, adverse events occurred in one-third of hospital admissions, even in hospitals that had instituted advanced patient safety programs. According to the WHO (http://www.who.int/patientsafety/en/), based on the incidence of adverse events worldwide, the chance of being harmed in healthcare is 1 in 10; 43 million of patient safety incidents occurred around the world; the risk is distinctively higher in developing countries.

Additionally, although the Evidence-based Clinical Practice Guidelines are aimed to improve quality of care for specific clinical cases, the adherence from the physicians is still insufficient [12]; the situation has a relevant impact in the quality of service in countries like United States [13] and is one of the factors which drive the low quality of healthcare in developing countries [14].

Aligned to the perspective of linking Evidence-based Medicine to Evidence-based Management to improve quality of care, Frolich [15] states a model on which the quality of care is defined according to the level in the healthcare system at which it is assessed.

The author presents Determinants of Quality of Care as features developed to improve quality of care, see Fig. 8.8.

Determinants can be implemented at one or more organizational levels: macro, meso, or micro level depending on the design of the determinant. See the following definitions proposed by the author:

1. *At the Macro level of countries and organizations,* Frolich defines quality of care based on frameworks with dimensions characterizing areas of care. National and large organizational frameworks assessing quality of care generally include measurements in the following dimensions:
   1. Access; 2. Effectiveness & Appropriateness; 3. Responsiveness;
   4. Safety; and 5. Equity.
2. *At the Meso level of organizations,* the quality of care and the definitions are more focused, the dimensions include: 1. Effectiveness of care;
   2. Compliance with clinical guidelines; 3. Patient-related quality (e.g., quality of life, patient satisfaction); and 4. Organizational Quality (e.g., safety, rate of re-hospitalization, average length of stay). The quality of care can also be defined in relation to specific technologies, such as care management practices.
3. The Micro level includes 1. Measures related to patients (quality of life, patient satisfaction); and 2. Providers (job satisfaction).

Fig. 8.9 summarizes the three organizational levels and their respective dimensions in the healthcare system.

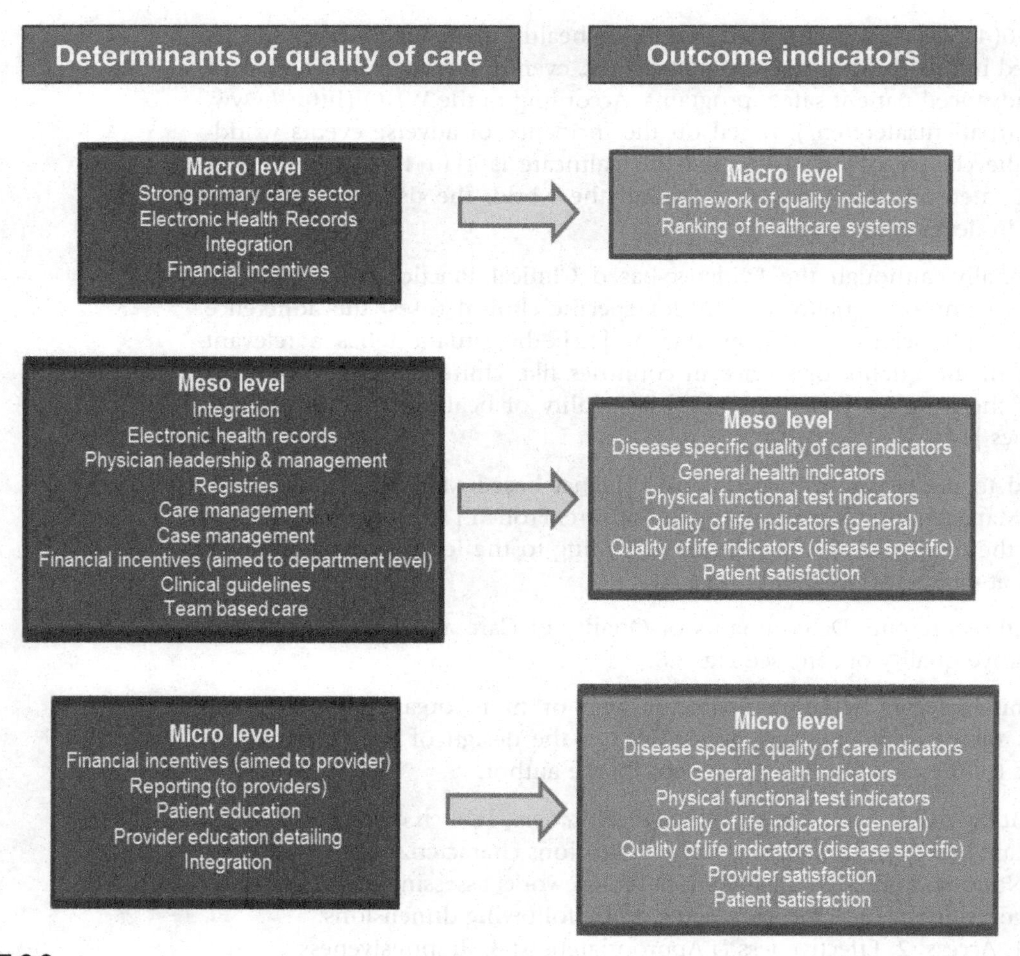

FIGURE 8.8

Determinants of Quality of Care and outcome indicators. *Frolich A. Identifying organizational principles and management practices important to the quality of health care services for chronic conditions. Dan Med J 2012;58(2):B4387.*

Some of the limitations of the management perspective are: the quality determinants at high-performing sites depend on the context; the mechanism of operation and the effect on quality of care of financial and disclosure incentives are complex and depend on shape, content, and design of the incentive.

The definition of the dimensions requires analyzing the health organization considering the strategic and operational aspects; certainly this framework contributes to the quality of the healthcare project.

Macro level

1. Access
2. Effectiveness & appropriateness
3. Responsiveness
4. Safety
5. Equity

Meso level

1. Effectiveness of care
2. Compliance with clinical guidelines
3. Patient-related quality (quality of life, patient satisfaction)
4. Organizational quality (safety, rate of re-hospitalization, average of stay)

Micro level

1. Measures related to patients (quality of life, patient satisfaction)
2. Measures related to providers (job satisfaction)

FIGURE 8.9
Dimensions of area of care. *Frolich (2012).*

8.3.3 Systems Approach to Improve Healthcare: Health-Related Technologies

Healthcare technology requires a systems approach as medical devices become connected to the Information Technology—IT network for interchanging data with the Electronic Health Records—EHR and other medical devices. In United States there is limited amount of appropriate curriculum and hands-on laboratory resources among the 87 US university-based BME programs to manage medical device interoperability [16].

Sloane [17] remarks that for the past decade, few medical devices were designed to operate in a vacuum. Most have one or more embedded computer and communication chips/modules that allow the devices to connect to other devices, hospital information systems—HIS, and/or specialized systems like Laboratory Information Systems—LIS and Radiology Information System—RIS.

Medical devices, HIS, LIS, and RIS products are now being designed to allow or even promote device-system integration and interoperability as the author states, as a consequence the devices must be safely and reliably perform their primary design function(s), but they also now send and receive data and patient information to other devices and the HIS. This context stablishes the need of an appropriate education, training, and credentials (http://www.who.int/ehealth/en/). Figs. 8.10 and 8.11 show the recommendations for Education and Training on Skills in Management and Leadership according to Sloane.

1. General Project Management, including Agile Methodologies.
2. Software and Systems Development Life Cycle (SDLC) Methodologies.
3. Software and System Engineering, including System of Systems Engineering, aka Complex Systems Engineering, including concepts of interdependencies, modeling and simulation, Software Quality Assurance, including Verification and Validation, including the "V-Model" process and system engineering approach, and Concurrent Engineering.
4. Human factors engineering, including human-system engineering.
5. Life cycle cost analysis (or Total Cost of Ownership for information and communication technologies—ICT).
6. Lean/six-sigma quality methods.
7. Risk management and risk mitigation (ISO 80001 et al.)
8. HTM Maturity Models, including the current HIMSS ERM Maturity Model.
9. Business process engineering/re-engineering and management of change.
10. Contract negotiation.

FIGURE 8.10
Management skills required improving healthcare. *Sloane E, Welsh J, Judd T. New opportunities for biomedical engineers—BE/Clinical Engineers—CE Health Information Technology Education, 2014.*

1. Recruitment, training, and retention of ICT professionals
2. Cross-training and team building for customer service
3. Job descriptions and careers that include analysts, trainers, implementers, etc.
4. Human capital management
5. Problem/conflict resolution, mediation

FIGURE 8.11
Leadership skills to improve healthcare. *Sloane E, Welsh J, Judd T. New opportunities for biomedical engineers—BE/Clinical Engineers—CE Health Information Technology Education, 2014.*

The System approach applied to Healthcare contributes to better understand and even to predict health needs also improve the elaboration of solutions. In the case of the increasing emerging Health-related technologies we observe an opportunity to improve patient outcomes; it is recommended though to consider the investment on several factors as education and training to be aligned to the requirements.

References

[1] Salinsky E, Gursky E. The case for transforming governmental public health. Health Affairs 2006;25(4):1017–2018. Available at: <http://content.healthaffairs.org/content/25/4/1017.full>.

[2] Frost & Sullivan Manufacturing Leadership Council. Top 10 Mega Trends, New York, 2012.

[3] SEBOK. Guide to the systems engineering body of knowledge (SeBoK). <http://sebokwiki.org/wiki/Guide_to_the_Systems_Engineering_Body_of_Knowledge_(SEBoK)>.

[4] Institute of Medicine. Crossing the quality chasm: a new health system for the 21st century. Washington, DC: National Academies Press; 2001.

[5] National Academy of Engineering, <https://www.nae.edu/>.

[6] Proctor P, Compton WD, Grossman J, Fanjiang G. Building a better delivery system: a new engineering/health care partnership, committee on engineering and the health care system. Washington DC: National Academy of Engineering and Institute of Medicine, National Academy Press; 2005.

[7] Kopach-Konrad R, Lawley M, Criswell M, et al. Applying systems engineering principles in improving health care delivery. J Gen Intern Med 2007;22(Suppl. 3):431–7. Available from: http://dx.doi.org/10.1007/s11606-007-0292-3.

[8] Ravitz A, et al. Systems approach and systems engineering applied to healthcare: improving patient safety and healthcare delivery. Johns Hopkins APL Tech Dig 2013;354 31(4).

[9] Lohr KN. Medicare: a strategy for quality assurance. Washington, DC: National Academies Press; 1990.

[10] World Health Organization—WHO. Quality and accreditation in healthcare services. A global review, Geneva, 2003.

[11] Dentzer S. Still crossing the quality chasm-or suspended over it?. Health Aff. 2011;30(4):554–5.

[12] Cabana MD, Rand CS, Powe NR, Wu AW, Wilson MH, Abboud PAC, et al. Why don't physicians follow clinical practice guidelines?: A framework for improvement. Jama 1999;282(15):1458–65.

[13] Kenefick H, Lee J, Fleishman V. Improving physician adherence to clinical practice guidelines: barriers and strategies for change. New England Healthcare Institute; Massachusetts, US, 2008.

[14] World Health Organization. Guidelines for WHO guidelines, Geneva, 2003.

[15] Frolich A. Identifying organizational principles and management practices important to the quality of health care services for chronic conditions. Dan Med J 2012;58(2):B4387.

[16] Morton A. UBT program: preparing the health IT leaders of tomorrow, today, HHS ONC. Retrieved from: US Government Office of the National Coordinator (ONC) for Health; 2011.

[17] Sloane E, Welsh J, Judd T. New opportunities for biomedical engineers—BE/Clinical Engineers—CE Health Information Technology Education, 2014.

CHAPTER 9

New Organizational Model for Hospitals in the New Technology Context

The main obstacle to hospital technological development is its own organizational structure. A new model is urgently required.

In these times, it is still assumed that hospitals should only maintain their medical equipment so that all technology works well and health professionals can care for their patients. This is the concept or model for more than 70 years and we should not think of another model given that this is apparently working well, and in places where it does not work well, it is not for the model but could be by the deficient application of this one. These premises may be irritating to some biomedical engineers or clinical engineers and at this point, we want to clarify the reasons. The concept of what we understand by "technology in health" in a hospital is not more than 15 years, so we are still in the process of transformation or at least adequacy. Thus, with regard to technology, we are moving from the concept of "managing assets" to the concept of "managing technology," or put another way, we are moving from the premise "maintenance of medical equipment" to the premise "functional clinical service." The idea that the maintenance of medical equipment is the only thing that we must attend as a technological component has led us to contradict ourselves. We all recognize the potential of new technological developments, such as nanotechnology, genetic treatments, information systems, telemedicine, and robotics, we value this potential, but on the other hand hospitals suffer of the capacity to assimilate these technologies, that is to say, this potential is neglected, except for some hospitals that, due to having units of research in science and technology, partially solve the problem. Normally, hospitals think about buying medical equipment, installing and maintaining them, but basically it is a technology transfer, but all this happens as an administrative process. Recall from Chapter 1, Healthcare Technology Management (HTM) & Healthcare Technology Assessment (HTA), health technologies are

1. clinical technologies: clinical procedures, medical equipment, medical materials, and medicines;

Healthcare Technology Management Systems. DOI: http://dx.doi.org/10.1016/B978-0-12-811431-5.00009-6
© 2017 Elsevier Inc. All rights reserved.

2. support technologies: infrastructure, hospital equipment, information and communication systems, and organization;
3. preventive technologies;
4. protection technologies;
5. promotional technologies;
6. environmental health technologies.

Does the hospital have the ability to systematically manage these technologies? We all recognize that the organizational structure of today's hospitals is a complex network of committees, departments, services, and staff. From the point of view of patients, they need to determine who is in charge, what services to expect from whom and when, with what results, and at what cost to them. The model for the organization of hospitals has historically been taken from nonprofit community hospitals. The management, control, and governance of the hospital is divided into three influential entities: the medical staff, the administration, and the governing council [1]. The main operational divisions of the hospital represent areas of the hospital's functions, such as medical units, nursing, diagnosis, patient therapy, inspection, human resources, hotel services, and community relations. We will review the current organization of hospitals to inscribe there the meaning of a new model of organization that proposes to the hospital to have the ability to manage its own technology.

The medical staff is represented by a president or chief and acts as liaison between the hospital administration and medical staff members. The primary role of the medical staff is to recommend to the board the positions of medical professionals and provide supervision and peer review of the healthcare quality in hospitals. These functions are carried out through committees that interact with the administration and other committees of the director board. Complementing the medical staff are the doctors who completed their training and spent a period of practice, they are referred to as attending physicians, and the postgraduate doctors under the supervision of the care physicians are the so-called residents as they rotate to cover attention 24 hours. There is no universal rule of how the medical divisions or departments of the hospital can be organized; often the hospital purposes and the medical staff's specialty determine the divisions. The main departments are internal medicine, surgery, obstetrics and gynecology, and pediatrics. In large hospitals and teaching hospitals departments of internal medicine can be divided into subspecialties such as cardiology, ophthalmology, urology, oncology, gastroenterology, pulmonary medicine, endocrinology, otolaryngology, and many others. Surgery subspecialties may include orthopedic surgery, thoracic surgery, neurosurgery, cardiac surgery, plastic and reconstructive surgery, among others. Each division or department is headed by a medical chief who is in charge of overseeing the practice and quality of service delivered.

On the other hand we have the nursing unit, which comprises one of the largest components of the hospital organization. It is divided according to the type of patient care to be delivered between the different medical specialties. Nursing units are composed of a patient beds number grouped within medical areas to allow the centralization of special facilities, supplies, equipment, and personnel appropriate to the needs of each patient type. For example, the equipment type, skills, and level of patient requirements vary considerably between an orthopedic unit and an intensive care unit. The head nurse has responsibility for all nursing care in your unit. Such care includes conducting medication orders from doctors and medical residents, diets, and different types of therapies. The head nurse also coordinates all types of patient care which may include services by other units of the hospital, such as the diets department, physical therapy, pharmacy, or clinical laboratory. This work requires permanent attention of 24 hours, therefore, schedules are scheduled for 8 hours.

In addition, allied health professionals, such as technologists and technicians in clinical laboratory, physical therapy, occupational therapy, behavioral science professionals, and specialists for the support of service personnel such as nutritionists, health information, clinical record assistants, among others.

The administrative branch of a typical hospital consists of a wide variety of nonmedical services which are essential for managing the physical plant and economic services of the hospital. Patients interact with two of them: the admission department and the payment office through which the hospital stay ends. The general administration is headed by a chief who is responsible for the day-to-day management of the hospital and is responsible for supervising the administrative departments in charge of financial operations, public relations and personnel, logistics, among others. Large hospitals have a chief operating officer in each of these departments. The administrative heads usually conform the level of vice-directors of the hospital. In recent years, new departments have been created, one of which is the information systems management. This department is usually led by computer science professionals and is responsible for hospital information systems and data processing such as recording medical records or transcribing the services provided to convert them into financial information or processing patients' data to evaluate quality and cost effectiveness of the hospital services. Likewise, the administrative branch is responsible for hotel services such as building maintenance, security, laundry, television, and telephone services.

Types of hospitals: acute-care hospitals have an average stay time of less than 30 days and can be operated by nonprofit entities, managed by for-profit corporations, and public hospitals operated and managed by government entities. Hospitals can also be classified as teaching hospitals affiliated with medical schools and nonteaching hospitals. Both can provide clinical training for nurses, allied health personnel, and a wide variety of health technicians. In some countries there is a classification according to the complexity of the health services offered, so the most complex hospitals are from levels III-3, III-2, III-1, and so on up to level I-1, followed by small establishment health services such as medical posts. Lower level hospitals and medical posts comprise primary healthcare. Typically, the largest and most complex hospital in the region is the head of a health network. It is assumed that patients should go to lower level hospitals for the first time and then scale according to the complexity of the clinical requirement, however, in practice it does not work exactly as well. Likewise, health sector reforms are underway to ensure universal health, as well as to ensure the hospital organization autonomy, parastatal corporatization to confront the market, and privatization in order to disassociate with entities rectors of public sector [2] with the purpose of guaranteeing the quality, coverage, and opportunity of health services. However, even today, every health system is in a process of change and adaptation.

Fig. 9.1 shows an example of a traditional hospital organization chart. It can be seen that the ability to handle various types of technology is not obvious or even considered. One possible explanation is that the hospitals organization result of a experiences learned series along the way according to the different circumstances of great importance, and being undoubtedly a technological entity in which it has become today, the hospital needs to rethink its organization structure in the current scenario in which the presence of technology is evident and needs to be managed. Another explanation is the understanding, sometimes confusing, about technical and technology by managers and some health professionals. Under the concept of technical, the technological aspects are reduced to objects or assets that need to be operative through maintenance actions, leaving aside technology own actions as they are:

- Analysis capacity for planning
- Healthcare technology assessment
- Capacity to interpret the impacts of technology in the people's health
- Cost control according to the possibilities that the technology offers
- Ability to design or adapt technology to the hospital's own needs
- Risk analysis and alternative approaches for the greatest benefit of technology in people
- Other actions derived from the concept of technology in health

We approach the issue from the technology side; however, we think that many other medical and administrative aspects are also pending. In this sense, global governing bodies opt for autonomy in managing hospitals so

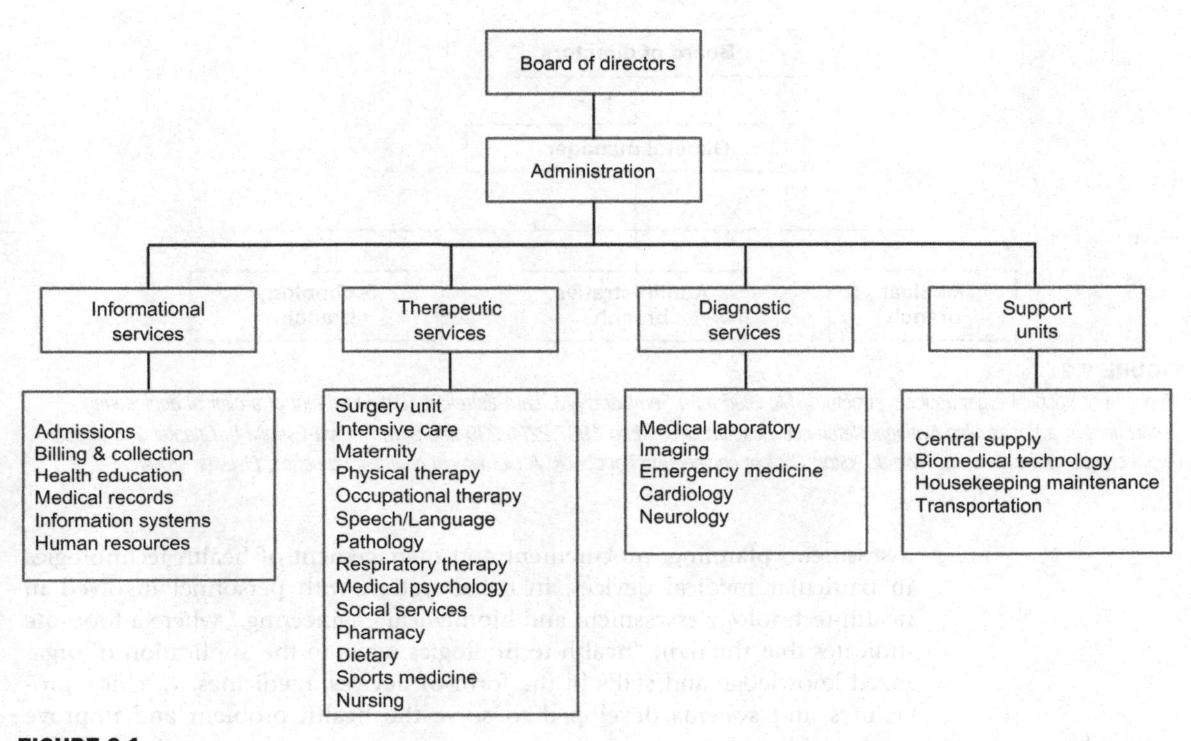

FIGURE 9.1
Example of a traditional hospital organizational chart.

that they can be adapted to current circumstances. Particularly in the case of technology we have important aspects that we have considered in the previous chapters that will now lead us to propose alternatives of organizational structures more conveniently to meet hospital purposes and for patients benefit.

"Under the traditional model of organizational structure of hospitals, technology management does not exist, consequently decisions concerning technology are not normally considered within the hospital management" [3,4]. Because of this, the authors propose to have a branch of technology in the hospital organizational structure based on the experiences of pilot projects carried out in Venezuela in the 1990s, which they called Department of Clinical Engineering, however, given the current definitions of health technology can be considered in generic terms as a branch of technology as shown in Fig. 9.2. Most recently, the World Health Assembly in its resolution WHA 60.29, 2007 [5] urges to member states "to formulate appropriate national strategies and plans for the establishment of systems for the

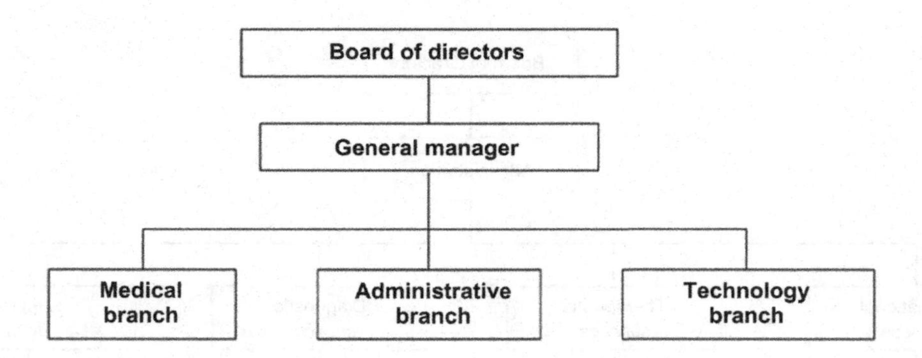

FIGURE 9.2
Proposed of hospital organizational structure. *Modified from Seminario RM, Lara-Estrella L. Establishment of a clinical engineering department in a Venezuelan National Reference Hospital. J Clin Eng 1997;22(4):239 and Silva R, Lara-Estrella L. Chapter 26: Clinical engineering in Venezuela. In: Dyro J, editor. Clinical engineering handbook. A biomedical engineering series. Elsevier; 2004.*

assessment, planning, procurement and management of health technologies in particular medical devices, in collaboration with personnel involved in health-technology assessment and biomedical engineering," where a footnote indicates that the term "health technologies refers to the application of organized knowledge and skills in the form of devices, medicines, vaccines, procedures and systems developed to solve the health problem and improve quality of lives."

The proposal in Fig. 9.2 summarizes a strategy that we will analyze according to its concept, its impacts, and its viability in the health systems. First, in organizational structures, divisions or branches do not mean separation of components for each to take its course, but is a way of grouping human resources, capacities, functions, processes, and similar responsibilities, but in such a way that allows the interaction between the different branches or divisions. In fact the interaction between the current medical branch and the administrative branch is habitual in the day to day. On the other hand, the current organization of hospitals usually has the department or office of engineering or maintenance, the only entity formally linked to technology, normally incorporated in the administrative branch as a technical support unit, which causes much of the obvious problems with respect to technology. The technical support and maintenance of medical equipment are very necessary and should be specialized as biomedical technicians do in different categories, however, given the level of complexity of technology in hospitals it is necessary to have capacities based on scientific knowledge, oriented to analysis, design, management, and self-learning skills, communication and negotiation in the fields of health technology, which means that the hospital does not only require technical support, but also has a capacity of professional management of technology to propose and direct the planning, acquisition, technological resources management, risk management, human resource development, and applied research needed to have functional clinical services from the point of view of the technology. We consider that HTM is not feasible from the administrative branch because the roles and functions are different, the first one has the role of financial operation, public and personnel relationship, logistical management of assets and liabilities, legal actions, i.e., carry out the "business" of the hospital, instead the second would be responsible for the clinically effective, efficient, safe, and cost-effective technology functionality in clinical services. These two responsibilities are completely different and concern equally different professional profiles. For the administrative branch the related professional profiles are administrators, accountants, statisticians, computer scientists, among others, but for the technology branch the professional profile points to biomedical engineers or clinical engineers, or other types of engineers but who have a formation specific for health.

How is the technology branch defined in the organizational structure of a hospital? We consider that the mission and vision of this technology branch can be exposed as follows:

Mission: To ensure the technological environment functionality of clinical services at the hospital, taking into account the different types of health technology, based on high levels of clinical effectiveness, efficiency in the use of resources, safety in the application of technology, and control of costs derived from the use of technology, being its main processes planning, acquisition, technological asset management, risk management, human resource development, and applied research for the continuous improvement of the hospital technological environment.

Vision: Within the framework of the hospital mission and vision, contribute in guiding the evolution of the hospital healthcare for a highly specialized attention and quality to the population attached, with leadership in the excellence of the technological environment.

We propose here a paradigm shift regarding hospital management, and although we know that organizational change implies a complicated process, we also understand that for many hospitals this process is urgent and immediate. In other cases it can be considered a process of gradual change. In the latter case, a viable strategy is to introduce an incubation process, in which the highest management unit of the hospital ascribes the new technology unit, and this unit, step by step, will begin to assume the functions for the technology management in the hospital. Subsequently it is expected that this unit already implemented will end up forming the technology branch of the hospital. An experience of this type is found in [5]. From both experiences [4,5], it follows that the least prudent to do is to evolve from the current engineering department or maintenance department, since naturally there would be two major constraints, that coming from the administrative unit that tend to oppose all such changes and the restriction itself of the current maintenance staff whose, before these proposed changes, after many years of doing the maintenance work, will not always be willing to assimilate a new strategy. In any scenario, hospital managers must be very clear about their purposes and there is nothing better than having an institutional policy, whether local or regional that helps to establish this new branch in the organizational structure of the hospital.

In the next sections, we will describe alternatives for how to implement the functions and processes of the comprehensive HTM system to be carried out from the technology branch of hospital, as well as the professional profiles and resources required for this new organizational strategy.

9.1 FUNCTIONS AND PROCESSES FOR NEW INTEGRATED HEALTHCARE TECHNOLOGY MANAGEMENT SYSTEM

Organizations need to manage their activities and resources in order to guide them to achieve their objectives and institutional mission and vision. In the health sector, it is undeniable that hospitals are immersed in competitive and even globalized environments and markets, so it is not just about subsisting, but about succeeding in their purposes and having "good results." For this, hospitals need to configure their management system based on recognized standards or models. ISO 9000 defines a management system as one that allows establishing the policy and objectives, and then achieving those objectives. The European Foundation of Quality Management (EFQM) model defines a management system as a general scheme of processes and procedures that is used to ensure that the organization performs all the tasks necessary to achieve its objectives. Both models promote the adoption of a process-based approach in the management system as a basic principle to efficiently obtain the results related to customer and other stakeholder satisfaction. Under these considerations, the EFQM model has eight generic concepts that provide the theoretical guidelines that should guide the organization. These fundamental concepts are results orientation, customer focus, leadership and constancy of purpose, management by processes and facts, people development and involvement, continuous learning, improvement and innovation, partnership development, and corporate social responsibility. Regarding the structure and the contents of the model, the EFQM has nine criteria grouped in "enabler" and "result" criteria: the enabler criteria are

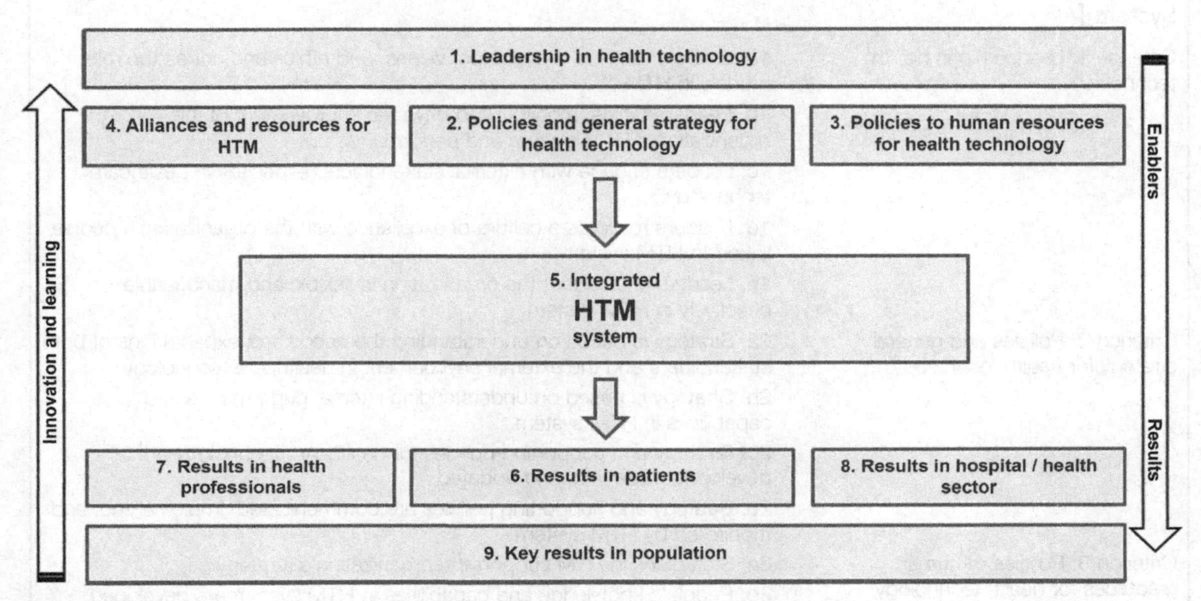

FIGURE 9.3
Conceptual model or criteria for an integrated Healthcare Technology Management (HTM) system for hospitals.

concerned with how the organization undertakes the key activities (leadership, policy and strategy, people, partnerships and Resources, and processes) and the result criteria are concerned with what results are being achieved (customer results, people results, society results, and key performance results) [6]. Guided by this framework of continuous quality improvement we propose a conceptual framework for the implementation of an integrated Health Technology Management system, as shown in Fig. 9.3.

Of course, it is possible to present many conceptual ways of hospital technology management systems, but this has the advantage of identifying what is evident in many hospitals, the lack of technology policies and the need to establish as an institutional goal the leadership in health technology. The processes of the integrated HTM system will be dealt with in the following paragraphs. Likewise, the model proposed in Fig. 9.3 describes the results of technology management in patients, but also in health professionals and in the hospital itself and in the health sector in general, of which every hospital is a part. The EFQM Introducing excellence [6] describes each of these nine criteria, which can be easily oriented to the purposes of an integrated Health Technology Management system (Table 9.1).

Table 9.1 EFQM Model Criteria Applied to Integrated Healthcare Technology Management (HTM) System [6]

Criterion 1: Leadership in health technology	1a. Leaders develop the mission, values, and ethics and act as the role models in HTM
	1b. Leaders define, monitor, and drive the improvement of the organization's HTM system and performance
	1c. Leaders engage with external stakeholders, especially in healthcare technology
	1d. Leaders reinforce a culture of excellence with the organization's people linked to HTM system
	1e. Leaders ensure that the organization is flexible and manageable effectively in HTM system
Criterion 2: Policies and general strategy for health technology	2a. Strategy is based on understanding the needs and expectations of both stakeholders and the external environment in healthcare technology
	2b. Strategy is based on understanding internal performance and capabilities in HTM system
	2c. Strategy and supporting policies for healthcare technology are developed, reviewed, and updated
	2d. Strategy and supporting policies are communicated, implemented, and monitored by HTM system
Criterion 3: Policies of human resources for health technology	3a. Staff plans in HTM support the organization's strategy
	3b. People's knowledge and capabilities in HTM system are developed
	3c. People are aligned, involved, and empowered in HTM
	3d. People of HTM system communicate effectively throughout the hospital organization
	3e. People are rewarded, cared for, and recognized by HTM system
Criterion 4: Alliances and resources for HTM	4a. Partners and suppliers of HTM system are managed for sustainable benefit
	4b. Finances of HTM system are managed to secure sustained success
	4c. Buildings, equipment, materials, and all HTM system resources are managed in a sustainable way
	4d. Healthcare technology is managed to support the delivery of its strategy
	4e. Information and knowledge are managed to support effective decision making and to build the HTM system capability
Criterion 5: Integrated HTM system	5a. HTM processes are designed and managed to optimize stakeholder value
	5b. Products and services of HTM system are developed to create optimum value for patients and users
	5c. Products and services of HTM system are effectively promoted and marketed
	5d. Products and services of HTM system are produced, delivered, and managed
	5e. Customer relationships of HTM system are managed and enhanced
Criterion 6: Results in patients	6a. Perceptions: These are the customer's perceptions of the HTM system. These perceptions should give a clear understanding of the effectiveness, from the customer's perspective, of the deployment and outcomes of the organization's customer strategy and supporting policies and processes
	6b. Performance indicators of HTM system: These are the internal measures used by the organization in order to monitor, understand, predict,

Continued

Table 9.1 EFQM Model Criteria Applied to Integrated Healthcare Technology Management (HTM) System [6] *Continued*

	and improve the performance of the organization and to predict their impact on the perceptions of their customers. These indicators should give a clear understanding of the deployment and impact of the organization's customer strategy, supporting policies, and processes
Criterion 7: Results in health professionals	7a. Perceptions: These are the people's perception of the HTM system. These perceptions should give a clear understanding of the effectiveness, from the people's perspective, of the deployment and outcomes of the organization's people strategy and supporting policies and processes
	7b. Performance indicators of HTM system: These are the internal measures used by the organization in order to monitor, understand, predict, and improve the performance of the organization's people and to predict their impact on perceptions
Criterion 8: Results in hospital/health sector	8a. Perceptions: This is Health Sector's perception of the HTM system. These perceptions should give a clear understanding of the effectiveness, from society's perspective, of the deployment and outcomes of the organization's societal and environmental strategy and supporting policies and processes
	8b. Performance indicators of HTM system: These are the internal measures used by the organization in order to monitor, understand, predict, and improve the performance of the organization and to predict their impact on the perceptions of the relevant stakeholder within health sector. These indicators should give a clear understanding of the deployment and impact of the organization's societal and environmental strategy and supporting policies and processes
Criterion 9: Key results	9a. High quality in health services for the benefit of patients, based on appropriate, functional and high availability of technology, this is high effectiveness in health services, high efficiency in the use of technological resources, maximum safety in the use of technology, and effective costs control about technology
	9b. Business outcomes of HTM system: These are the key financial and nonfinancial business outcomes that demonstrate the success of the organization's deployment of their strategy. The set of measures and relevant targets will be defined and agreed with the stakeholders
	9c. Business performance indicators of HTM system: These are the key financial and nonfinancial business indicators that are used to measure the organization's operational performance. They help monitor, understand, predict, and improve the organization's likely outcomes

With this holistic view, we also find compatible the application of the Performance Assessment Tool for quality improvement in Hospitals PATH [7], which provides specific orientations and indicators that in many cases affect the aspects of the technology. The PATH model is shown in Fig. 9.4. Then we present a summary of the dimensions and subdimensions considered for the performance evaluation of a hospital:

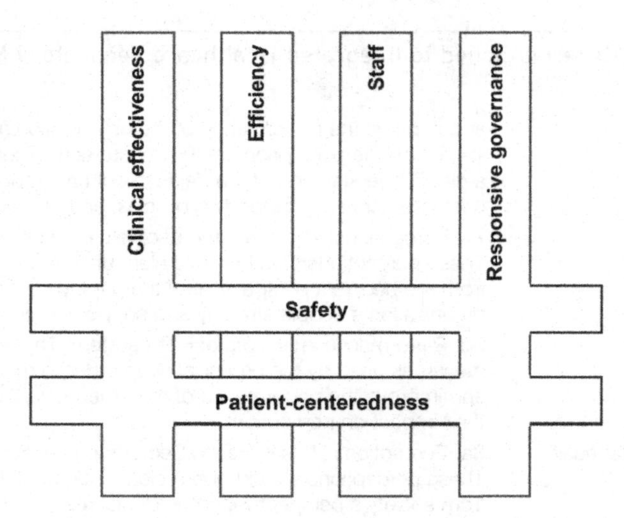

FIGURE 9.4
The PATH theoretical model to assess hospital performance [7].

Clinical effectiveness is a performance dimension, wherein a hospital, in line with the current state of knowledge, appropriately and competently delivers care or clinical services to, and achieves desired outcomes for all patients likely to benefit most. Subdimensions: Conformity of processes of care, outcomes of care processes, and appropriateness of care. The technological component is present in every clinical procedure, so it is part of the tasks of an integrated HTM system.

Efficiency is a hospital's optimal use of inputs to outputs maximal yield, given its available resources. Subdimensions: Appropriateness of services, input related to outputs of care, and use of available technology for best possible care. Technological resources must also be efficiently managed, which will complement the hospital's overall efficiency, and is therefore part of the objectives of the integrated HTM system.

Staff orientation and engagement is the degree to which hospital staff is appropriately qualified to deliver required patient care, has the opportunity for continued learning and training, works in positively enabling conditions, and is satisfied with their work. Subdimensions: Practice environment, perspectives and recognition of personal needs, health promotion activities and safety initiatives, behavioral responses, and health status. The technological component is present in the environment to clinical and administrative staff, so the integrated Health Technology Management system should ensure the functionality of this environment, including learning and training for the proper use of technology.

Responsive governance is the degree to which a hospital is responsive to community needs, ensures care continuity and coordination, promotes health, is innovative, and provides care to all citizens irrespective of racial, physical, cultural, social, demographic, or economic characteristics. Subdimensions: System/community integration, public health orientation. The ability to meet the health needs of the community is also given by the functional and technological resources available. This capability can be ensured through an integrated HTM system for hospitals.

Safety is the dimension of performance, wherein a hospital has the appropriate structure, and uses care delivery processes that measurably prevent or reduce harm or risk to patients, healthcare providers and the environment, and which also promote the notion. Subdimensions: Patient safety, staff safety, and environmental safety. With the clinical area being a technological space, it is necessary that this environment is also technologically safe for patients and hospital staff, and this dimension is particularly considered in an integrated HTM system technology.

Patient centeredness is a dimension of performance wherein a hospital places patients at the center of care and service delivery by paying particular attention to patients' and their families' needs, expectations, autonomy, access to hospital support networks, communication, confidentiality, dignity, choice of provider, and desire for prompt, timely care. Subdimensions: Client orientation, respect for patients. The best care to patients and families also has a technological environment whose functionality and security is crucial not only about having technology in operation but it is also functional for patients and users: This is one of the prerequisites for implementing a HTM system for hospitals.

What are the main processes of the integrated HTM system? In Fig. 9.5 is shown a general process map of HTM system with six main processes, which are described in more detail in the preceding chapters, but here describe the scope and the interaction between them to achieve the purpose of this HTM system, which are greater clinical effectiveness, greater efficiency in the use of technology, increased security due to application of technology, better cost control, and thus from the technological environment contribute to a higher quality of health services for the benefit of patients.

The integrated HTM system to hospitals shown in Fig. 9.5 can be explained by its inputs and outputs. Inputs are assumptions, policies, and institutional information to which an integrated HTM system responds to generate tangible results or outputs. The process map of integrated HTM system begins with two major sequential processes: planning and acquisition, and continues with two parallel processes: technological assets management and technology risk management. A complementary mental map was shown in

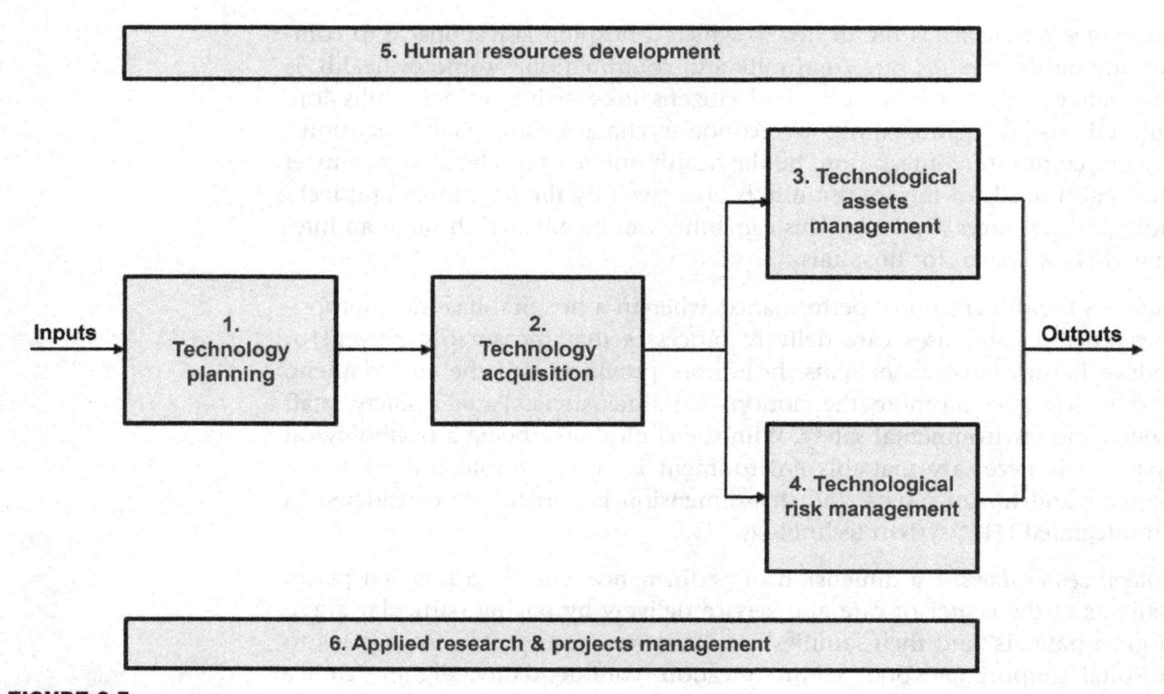

FIGURE 9.5
Main processes of an integrated Healthcare Technology Management (HTM) system for hospitals.

Fig. 1.5. In turn, all four operating processes mentioned are fed by two transverse processes: human resources development and applied research and projects management. Then the inputs and outputs of an integrated system of HTM are shown in Table 9.2.

Next, we will delineate the interaction between the processes of the integrated HTM system according to their particular inputs and outputs, in addition to the main internal subprocesses. At this point we consider the classification: Strategic processes, Operating processes, and Support processes.

1. *Planning process*:

   Planning activities in an integrated HTM system should in turn be strategically planned according to established institutional premises. Recall here that it is a question of considering all types of technology (clinical technologies, support technologies, prevention technologies, protection, promotion and environmental technologies). At least two types of operational processes are identified, the needs assessment for technological resources and economic evaluation. The outputs of the planning process are the planning reports which were mentioned in

Table 9.2 Inputs and Outputs of HTM Integrated System

Inputs of Integrated HTM System	Outputs of Integrated HTM System

• Institutional plans and policies: ISP Institutional Strategic Plan
• Clinical and administrative requirements related to technology in the hospital
• Situational status study of technology in the hospital, with indicators
• Plans and policies on health technology TS
  ◦ Historical record of annual operational plans related to health technology
  ◦ Historical record of indicators related to health technology
• Historical record of risk reduction plans in the hospital
• Historical record of plans for the development of human resources related to health technology
• Historical record of research and development plans on health technology
	High quality in health services for benefit of patients, based on appropriate, functional and high availability of technology:
	• High effectiveness in health services for benefit of patients
	• High efficiency in the use of technological resources
	• Maximum safety in the use of technology
	• Effective monitoring costs about technology

Chapter 4, Health Technology Planning and Acquisition. Fig. 9.6 shows a process map based on the ISO 9000 premises to explain the main subprocesses for planning of the integrated HTM system.

To carry out the planning work, it is necessary to have a wide variety of information products generated by it and the other HTM processes, such as inventory, nomenclature, needs certificate, management reports, contracts, indicators, infrastructure plans, training programs, investment projects, research projects, and budgets. We believe that for the management of all this information, the hospital must have a computerized and reliable information system, as well as methods for documentary management, which would be a valuable tool for all processes of the integrated HTM system. For this, the policy should be to feed the information system with each product and each update of the information.

2. *Acquisition process*:

The strategic process for the acquisition of technological products in health requires a strategic planning of the activities and resources necessary for the accomplishment of this work. It could not be said that this is an administrative task exclusively because it requires deep knowledge about the technologies, their characteristics, their health impacts, their costs, etc., as well as the interpretation of specialized information for their correct valuation in the decisions to be made.

At least two types of operational processes are identified, the preparation of the purchase and management of the proposals or

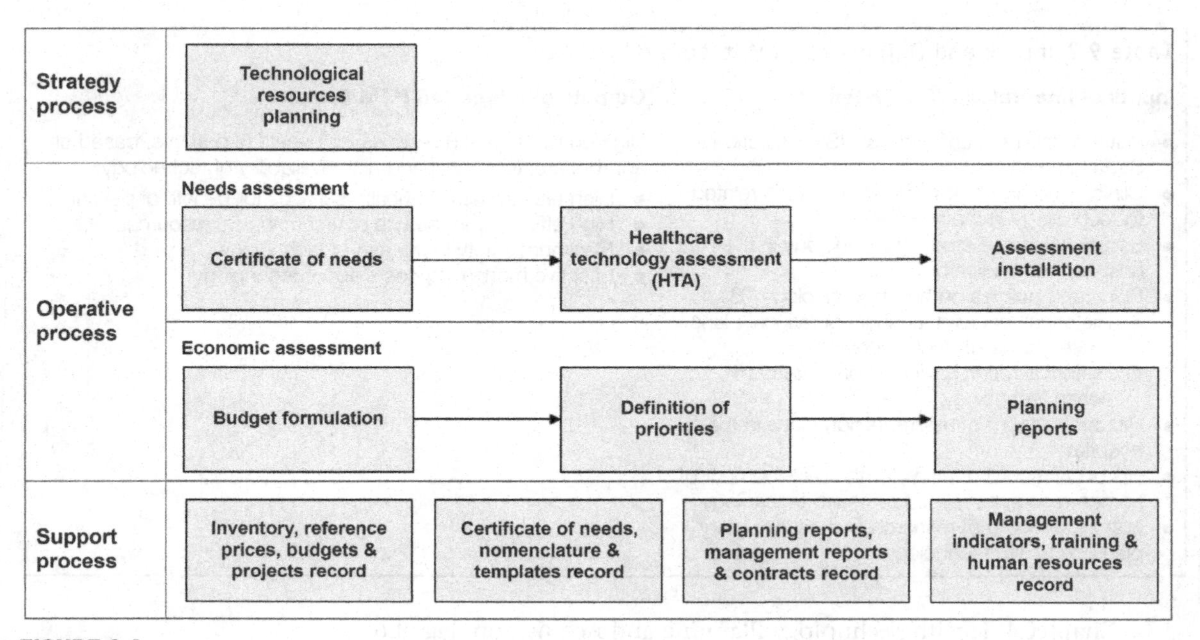

FIGURE 9.6
Main planning processes in the integrated HTM system for hospitals.

offers until the reception of the technological good purchased. The products of the procurement process are the technical specifications sheets, tender documents, evaluation report of the offers and the final purchase contracts. Further details on the products of the procurement process can be found in Chapter 5, Asset & Risk Management Related to Healthcare Technology. Fig. 9.7 shows a process map based on the ISO 9000 premises to explain the main subprocesses for the procurement of technology products to integrated HTM system.

3. *Healthcare Technological Assets Management*:

   The strategic process for the technological assets management in health should at least consider two subprocesses, the work planning to be carried out and the analysis to ensure the results. Given the complexity of the technological assets management, due to the number of items and the diversity in technological complexity, it is necessary to ensure the results in terms of effectiveness, efficiency, safety, and cost control. It is particularly well known that the current organizational structure model could be generating waste of the hospital's economic resources, so it is a matter of addressing this issue and reducing these wastes based on a more efficient management of technology, which

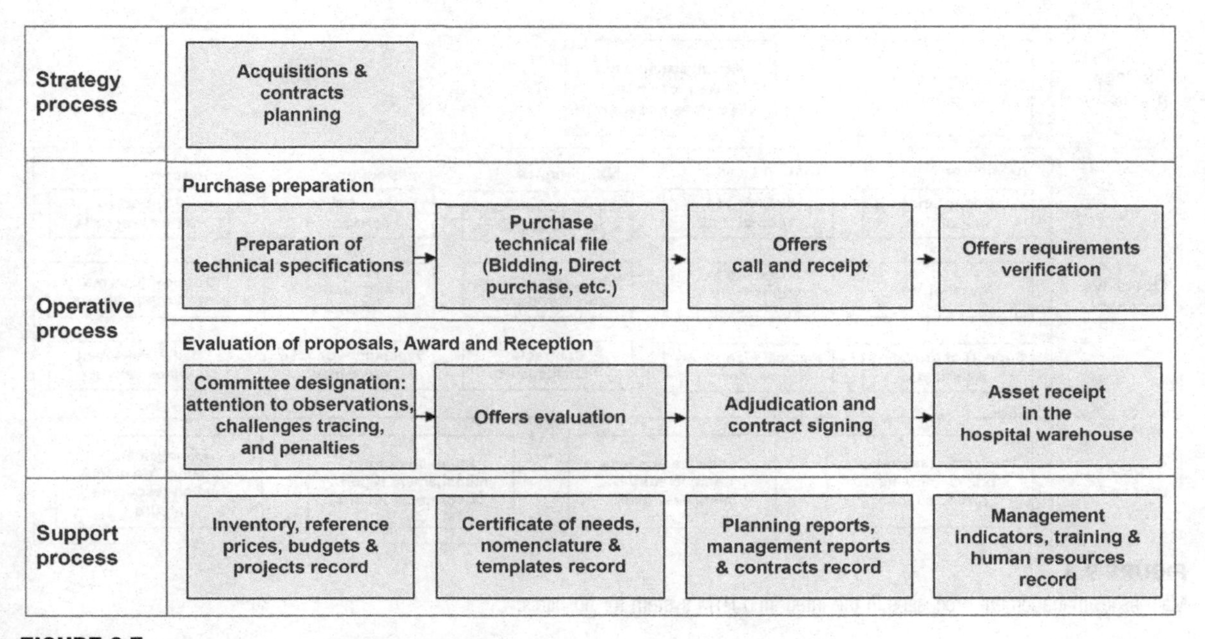

FIGURE 9.7
Main acquisition processes in the integrated HTM system for hospitals.

brings consequently, the savings achieved can be used to cover other expenses of the integrated HTM system.

The operational processes of the technological assets management are diverse, we have the installation of the purchased goods, the specific training or face-to-face to the users and staff of the integrated HTM system, corrective and preventive maintenance, the inspections described in Chapter 5, Asset & Risk Management Related to Healthcare Technology, and support from the technological approach required by all hospital units. Fig. 9.8 shows a process map based on the premises of ISO 9000 to explain the main subprocesses for the technological assets management of the integrated HTM system. These processes should be systematized to facilitate the high workload, as well as systematize the handling of information and documents.

4. *Risk Management process*:

The activities for risk management due to health technology should be planned as part of a strategic process to establish the achievements and resources necessary to carry out this work. The understanding of the impacts of risks and the interpretation of the rules and regulations require deep knowledge about the technologies and the impact on the

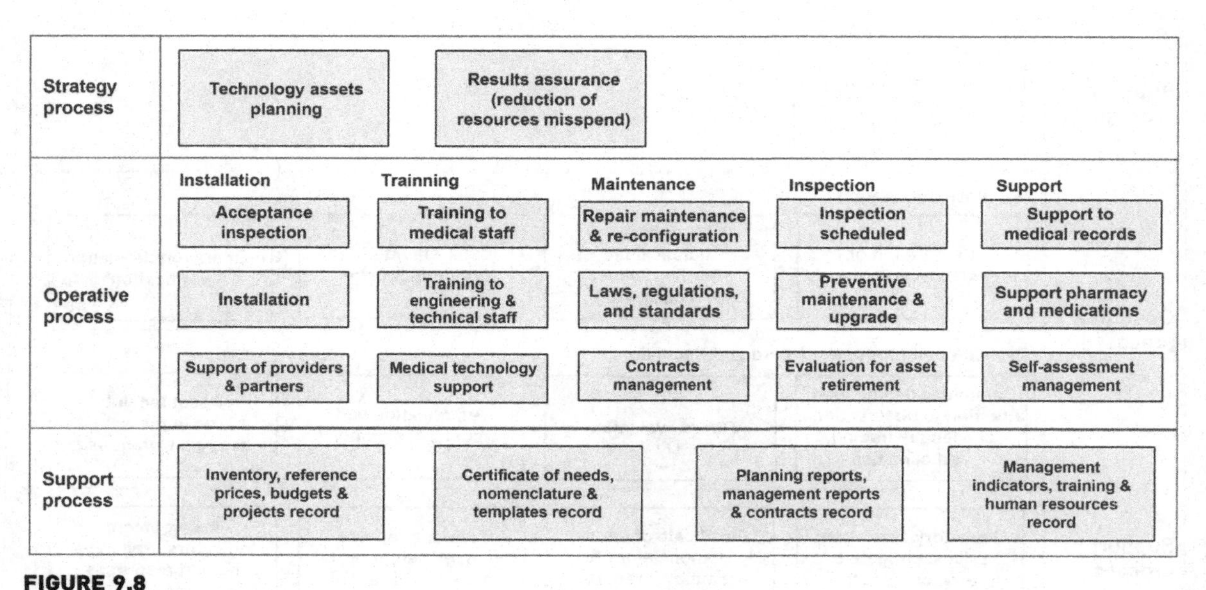

FIGURE 9.8
Main asset management processes in the integrated HTM system for hospitals.

physiology of the people for their correct valuation in the decisions to be made.

The operational processes include the tasks of techno surveillance, occupational health, metrological verification, infection control, hospital waste management, prevention and mitigation of disasters, and other aspects. Further details on the products of the risk management process can be found in Chapter 5, Asset & Risk Management Related to Healthcare Technology. Fig. 9.9 shows a process map based on the ISO 9000 premises to explain the major subprocesses for risk management due to the use of technology in the Integrated HTM system.

5. *Human Resources Development process*:

The development of human resources for the integrated HTM system should be planned as a strategic process according to the policies and institutional plans in order to propose the activities and resources necessary for the accomplishment of this work. The main target audiences are the medical-assistance staff and the own staff of the integrated HTM system. To define the training programs requires extensive experience and knowledge on teaching–learning methods, as well as deep knowledge about the technologies used in health.

The operational processes are oriented to create an appropriate environment for the development of the academic programs and the

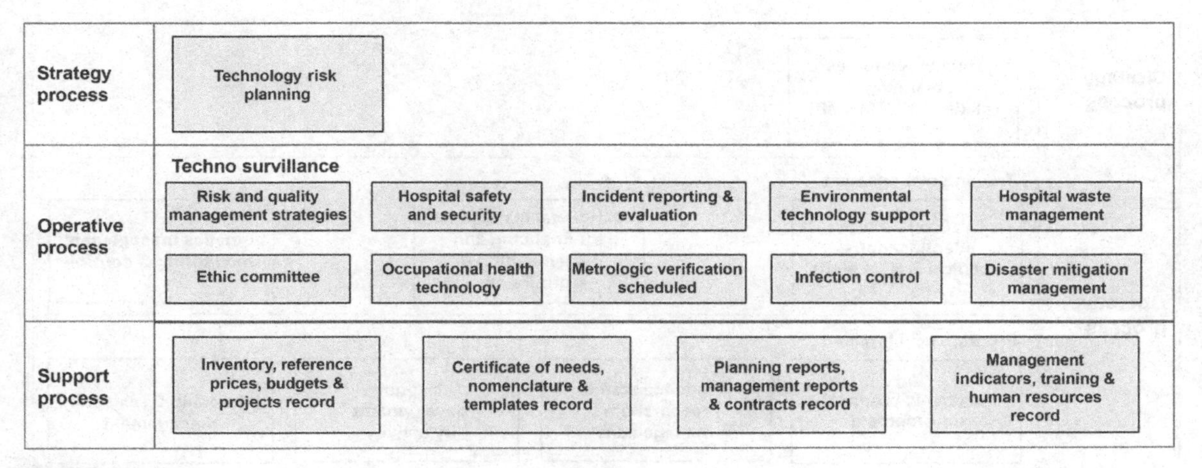

FIGURE 9.9
Main risk management processes in the integrated HTM system for hospitals.

execution of the programs themselves. One of the important tasks is to strengthen the interinstitutional relationship in order to establish alliances for the complement of capacities and also to facilitate the financing of the training programs. The products of the human resources development process are the reports of identification of requirements, interinstitutional agreements, academic proposals, budgets, the development of teaching capacities, and particularly the training of technical schools and universities students. Some more details about the products of the human resource development process can be seen in Chapter 6, Quality & Effectiveness Improvement in the Hospital: Achieving Sustained Outcomes. Fig. 9.10 shows a process map based on the ISO 9000 premises to explain the major subprocesses for the human resources development of the integrated HTM system.

6. *Applied Research and Project Management process*:

   The applied research and the management of the projects must be planned; this is done as a strategic process considering the institutional goals and objectives such that the activities and the necessary resources are identified. One of the goals of applied research is to improve HTM's own processes. Also, the evaluation of technology in health and the development of the technological components of clinical practice guidelines are requirements of a hospital with capacity to manage its technology. Another aspect to investigate and develop is the design of clinical services. Project development is another important process that requires formulating projects, seeking funding sources, executing the

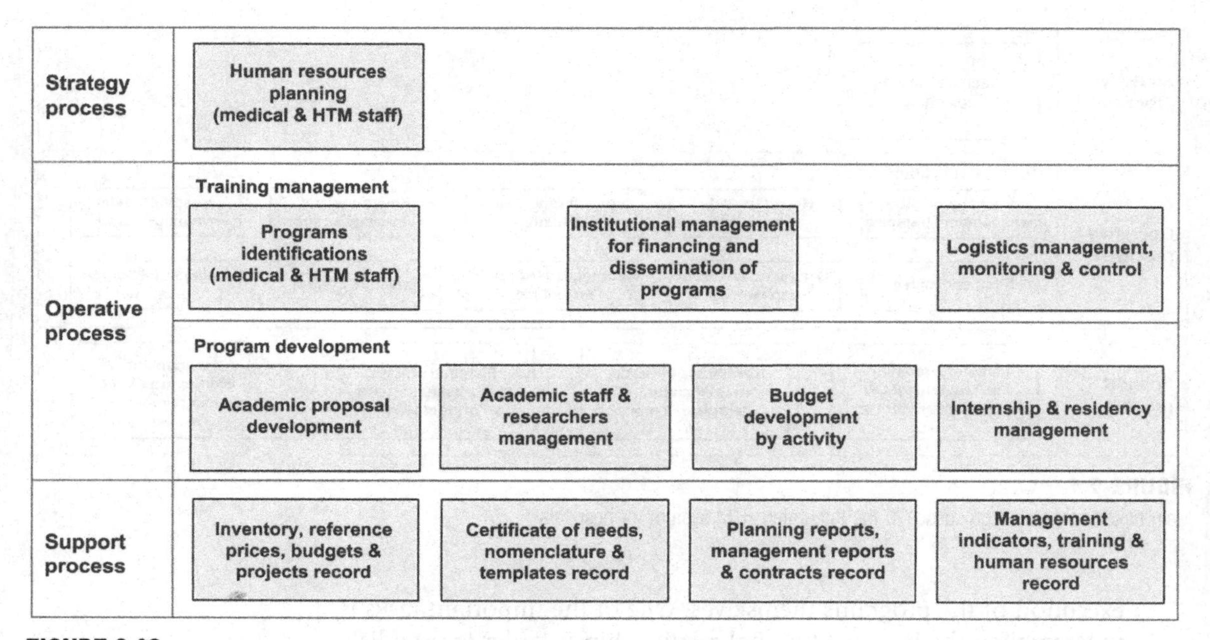

FIGURE 9.10
Main human resources management processes in the integrated HTM system for hospitals.

project, and finally entering the technology transfer, internally and externally to the hospital, in order to achieve economic returns wherever possible.

The products of the applied research process and projects development are publications, reports, and especially the implementation of the results in the improvement of the processes and status of the hospital itself. Some more details about the products can be seen in Chapter 7, Applied Research & Innovation in Healthcare Technology. Fig. 9.11 shows a process map based on the ISO 9000 premises to explain the main subprocesses for the applied research and project management of the integrated HTM system.

9.2 NEW ORGANIZATION STRUCTURE FOR NEW HEALTHCARE TECHNOLOGY MANAGEMENT SYSTEM: A PROPOSAL

We believe that the organizational structure of hospitals must evolve and adapt to the new context, particularly in the face of major technological advances in health, but also due to the great advances in medicine and

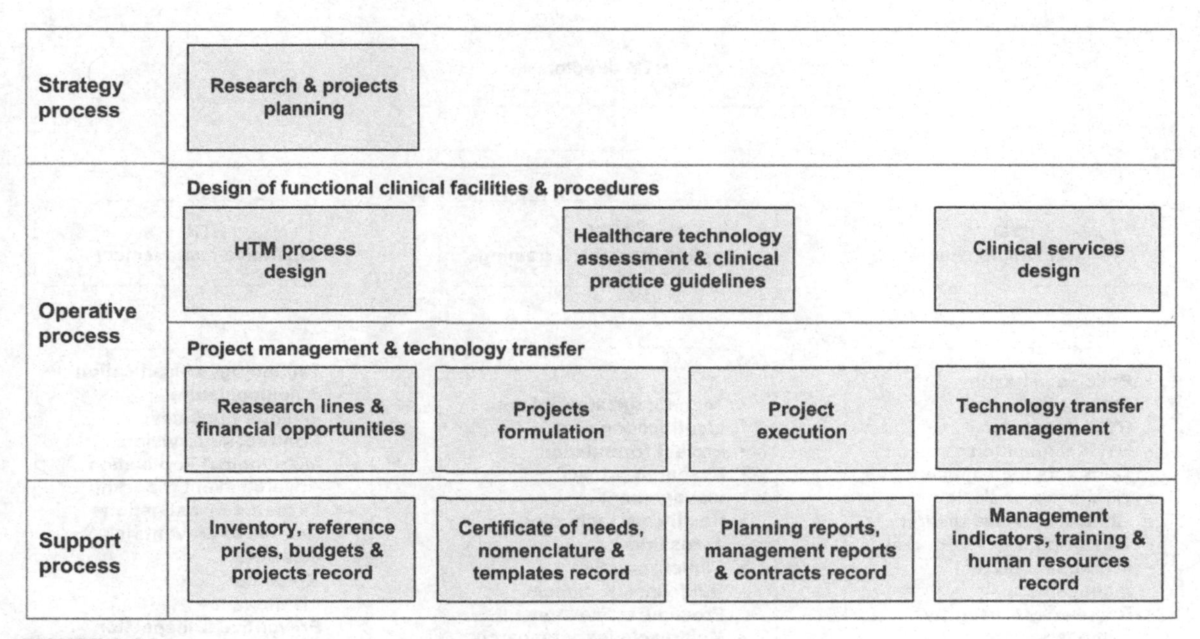

FIGURE 9.11
Main applied research and project management processes in the integrated HTM system for hospitals.

biology. In previous chapters we have gathered premises and approaches to what happens in the hospital on a daily basis that explain that a new model is necessary and even urgent to ensure the quality of health services. Next, we will present, as a proposal and exercise, a new organizational structure for hospitals, in particular as regards the technology branch, which should house integrated Health Technology Management system.

Fig. 9.12 summarizes the main functions of the integral HTM system. These have been grouped into three lines of work. One is the program management of the HTM from which the tasks of analysis and design are carried out. Another is the operational management from which the plans are executed with special emphasis in the assets management and risk management, and the third groups the functions to applied research, projects, and human resources development due to the ability of researchers to guide professional development and training activities. Of course it is possible to find another way to group the functions, however, we consider that this proposal will allow a better way to define the professional profiles, and therefore, will create a better environment for professional development.

Placing in the case of greater complexity such as large hospitals, head of health network, or reference hospitals, charges can be distributed as shown

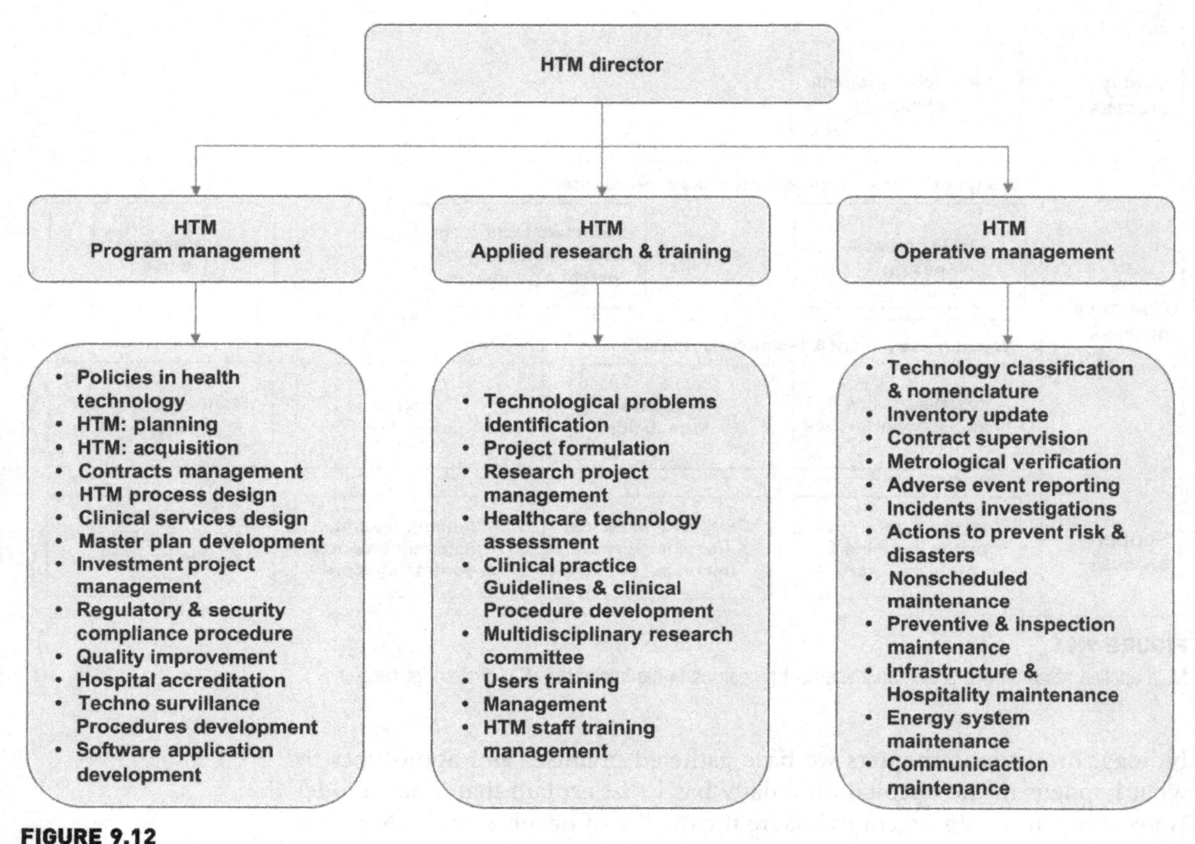

FIGURE 9.12
Technology branch: main functions in the integrated Healthcare Technology Management (HTM) system for hospitals.

in Fig. 9.13. This organization chart may grow or be smaller depending on the type of hospital, However we propose to maintain the three lines of work mentioned. This should include a director of the integrated HTM system, which is managed and supervised by a Board of Directors composed of members of HTM but also representatives of other units of the hospital, particularly senior officials of the medical branch and the administrative branch. Each line of work must be in charge of a director, who in turn will lead the working group in charge. Recall that these clusters are also due to the distribution of responsibilities and should not be considered isolated from each other, conversely, it is expected a great multidisciplinary interaction between each line of work, and also with other units of the hospital.

Finally we must emphasize that an integrated HTM system is by definition multidisciplinary and therefore it is hoped to incorporate diverse

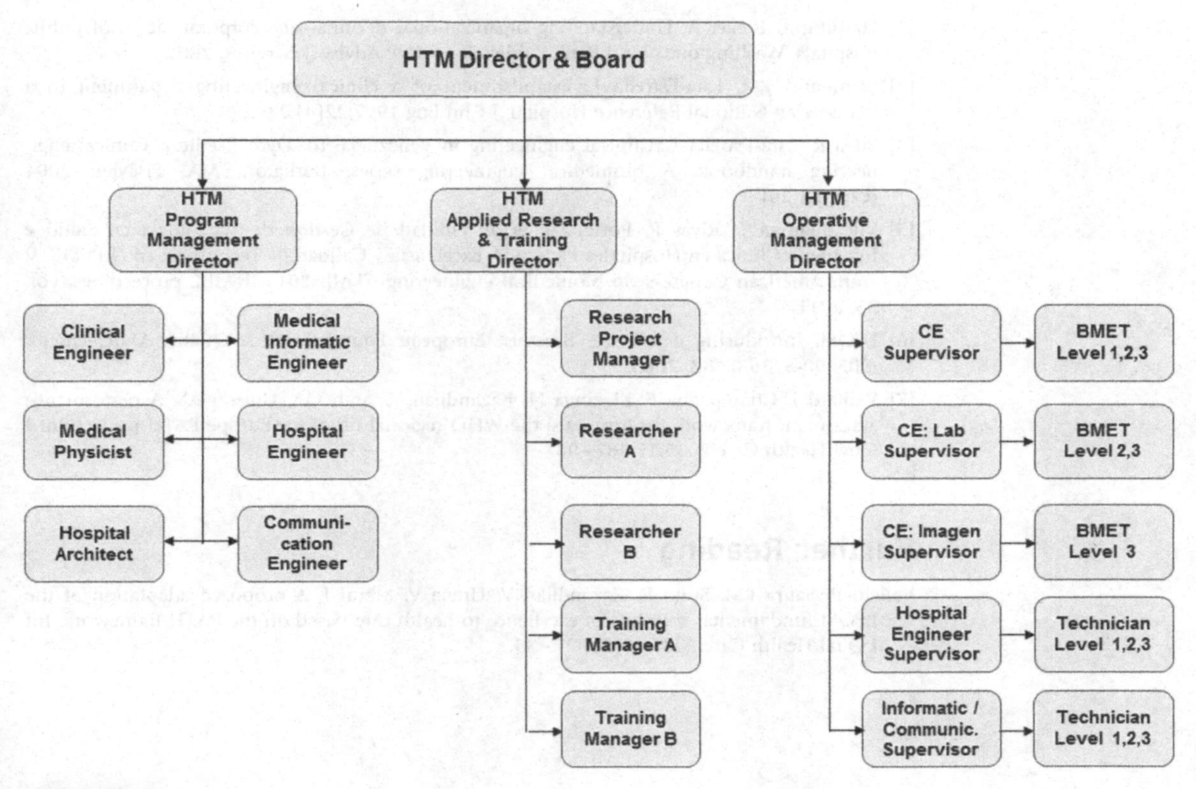

FIGURE 9.13
Technology branch: example of organigram to large hospital in the integrated Healthcare Technology Management (HTM) system for hospitals.

professional profiles and training such as biomedical engineers, clinical engineers, biomedical technicians, also hospital engineers, medical physicists, hospital architects, among other science, and engineering professionals. We believe that this new model of hospital organizational structure, holistically referred to as technology, will allow us to address the current problems of technology in hospitals, but it will also allow us to consider future prospects for hospitals, which require a greater capacity to assimilate new technology for the benefit of patients through a better quality of health services.

References

[1] Sultz H, Young K. Health care USA, understanding its organization and delivery, 7th ed. Burlington: Jones & Bartlett Learning; 2011.

[2] Harding A, Preker A. Understanding organizational reforms—the corporatization of public hospitals. Washington: World Bank Publication. HNP Advisory Services; 2000.

[3] Seminario RM, Lara-Estrella L. Establishment of a clinical engineering department in a Venezuelan National Reference Hospital. J Clin Eng 1997;22(4):239.

[4] Silva R, Lara-Estrella L. Clinical engineering in Venezuela. In: Dyro J, editor. Clinical engineering handbook. A biomedical engineering series. Burligton, MA: Elsevier; 2004 [Chapter 26].

[5] Vilcahuamán L, Rivas R, Portella J, et al. Unidad de Gestión de Tecnología en Salud e Ingeniería Clínica en Hospitales Peruanos: Excelencia y Calidad de Tecnología en el INMP. V Latin American Congress on biomedical engineering CLAIB 2011. IFMBE proceedings, vol. 33, 2011.

[6] EFQM. Introducing excellence. Brussels: European Foundation for Quality Management. ISBN 90-5236-072-3; 2003.

[7] Veillard J, Champagne F, Klazinga N, Kazandjian V, Arah OA, Guisset AL. A performance assessment framework for hospitals: the WHO regional office for Europe PATH project. Int J Qual Health Care 2005;17:487–96.

Further Reading

Vallejo P, Saura RM, Suno R, Kazandjian V, Ureña V, Mauri J. A proposed adaptation of the EFQM fundamental concepts of excellence to health care based on the PATH framework. Int J Qual Health Care 2006;18(5):327–35.

Index

Note: Page numbers followed by "*f*" and "*t*" refer to figures and tables, respectively.

Printed in the United States
By Bookmasters